有机化学实验
（Experimental Organic Chemistry）
（英汉双语版）

（供基础、临床、预防、口腔、护理、检验、影像及麻醉等专业用）

郭今心　朱荣秀　主编

山东大学出版社
SHANDONG UNIVERSITY PRESS
·济南·

内 容 简 介

《有机化学实验(英汉双语版)》根据我国高等医学院校双语教学及全英文教学的需要编写而成,全书内容分六部分,包括有机化学实验的基础知识、有机化学基础实验、有机化合物的制备、综合性和设计性实验、文献实验以及附录。在基础实验和有机化合物的制备实验部分,本书增加了废物处理的内容以提高学生的环保意识。结合学科发展,为开拓学生的视野,本书还增加了计算机分子模拟实验、知识拓展以及文献实验等内容。本书内容精炼,中英文对照编写,部分内容配有教学微课和实验视频,读者可扫描二维码学习。本书不仅适合医学院校本科生和留学生使用,而且可作为生物、环境等相关专业科研人员的参考用书。

图书在版编目(CIP)数据

有机化学实验:英、汉/郭今心,朱荣秀主编.—济南:山东大学出版社,2022.2(2024.3 重印)

ISBN 978-7-5607-7393-3

Ⅰ.①有… Ⅱ.①郭… ②朱… Ⅲ.①有机化学—化学实验—医学院校—教材—英、汉 Ⅳ.①062-33

中国版本图书馆 CIP 数据核字(2022)第 033957 号

责任编辑 祝清亮
文案编辑 曲文蕾 孙艳凤 任梦
封面设计 王秋忆

出版发行 山东大学出版社
社　　址 山东省济南市山大南路 20 号
邮政编码 250100
发行热线 (0531)88363008
经　　销 新华书店
印　　刷 济南乾丰云印刷科技有限公司
规　　格 787 毫米×1092 毫米 1/16
　　　　 13 印张 300 千字
版　　次 2022 年 2 月第 1 版
印　　次 2024 年 3 月第 2 次印刷
定　　价 39.80 元

《有机化学实验(英汉双语版)》编委会

主　编　郭今心　朱荣秀

副主编　唐龙骞　张永强　邢鹏遥

编　者　(按姓氏笔画排序)

邢鹏遥　朱荣秀　刘　刚　江　杰　孙继超　李明霞

邱晓勇　张　瑾　张永强　郭今心　唐龙骞

前言

本书在庞华、郭今心主编的《有机化学实验》基础上，根据我国医学教育综合改革的要求，配合国家“新医科”建设，针对医学院校学生的特点，参考国内外多种同类教材的内容，经多次修改编写而成。本书为中英双语版，主要面向医学院校基础、临床、预防、口腔、护理等专业的有机化学实验双语教学以及来华留学生的全英文教学。

本书以学生为中心，在“知识传授、能力培养、素质提高、协调发展”的教育理念下，构建了有机化学实验的基础知识、有机化学基本实验、有机化合物的制备实验、综合性和设计性实验以及文献实验等循序渐进的实验教学章节。

(1)第一章重点强调了实验安全知识，对事故的预防和处理进行了详细阐述，特别强调了有机化学实验废物的处理问题，并介绍了绿色化学的十二条原则，希望以此提高学生的安全意识和环保意识。为了让学生了解有机化学实验技术的最新进展，本书还增加了有机化学实验的新技术和新方法的内容介绍，同时还介绍了化学学科重要的文献资料以及搜索引擎和数据库的使用方法。

(2)在第二、三章中，为体现绿色化学的时代特色和减少环境污染，实验尽量以小量和半微量为主，尽可能使用无毒和低毒的药品试剂，且增加了废物处理内容。

(3)第四章有助于锻炼学生的学科思维能力，检验学生的综合实践能力，使学生能够用科学的思维模式、方法来分析和解决问题，实现“知行合一”。此外，由于计算化学在有机化学领域的应用日益广泛和重要，其对有机合成和现代药物设计具有非常重要的指导意义，因此本书还增加了计算机分子模拟实验。

(4)针对实验学时的限制和医学院校学生的专业特点，本书增加了视觉的化学、镇痛药等与医药有关的知识拓展以及文献实验。

本书利用现代化信息技术，将知识拓展、有机化学的文献资料及部分实验内容与二维码技术相结合，支持纸质教材与视频、微课融为一体的多元化教学手段。

本书主要由山东大学化学与化工学院的老师编写，编写分工如下：郭今心负责全书的编排与全书的修订和统稿,并编写 1.1～1.8 节、实验 2.9.2 以及实验 2.12;朱荣秀负责部分审稿和修改工作,并编写实验 2.6、实验 4.4、实验 4.8 以及文献实验;张永强负责编写 1.9 节、实验 4.1、实验 4.2、实验 4.6、实验 4.7 以及附录部分;刑鹏遙负责编写实验 3.3～实验 3.5、实验 3.7 以及实验 3.8;刘刚负责编写实验 2.9.1、实验 2.9.3 以及实验 4.3;孙继超负责编写实验 2.4、实验 2.7 以及实验 2.8;邱晓勇负责编写实验 2.11;江杰负责编写实验 2.1、实验 2.2 以及实验 2.5;张瑾负责编写实验 2.10 并绘制了第一章的实验仪器和装置图;李明霞负责编写实验 2.1、实验 2.5;山东大学药学院的唐龙骞负责部分修改工作,并编写实验 2.13、实验 3.1、实验 3.2、实验 3.6 以及实验 3.9。

本书是一本由师生共同打造的立体化教材,促进了教学相长,也体现了我们“教”与“学”相互融合的教学理念。实验微课分工如下:张瑾负责实验基础和安全教育,吕青负责实验 2.1,江杰负责实验 2.2,王舒平负责实验 2.9.1,王挺负责实验 2.9.2,吕香英负责实验 2 .10,郭今心负责实验 2.12,王旭负责实验 3 .8,李磊负责实验 4.1,刘刚负责实验 4 .3,朱荣秀负责实验 4.7。

实验操作视频由山东大学齐鲁医学院陈智言、林宗萱、吴念孜、唐嘉甫、王天姿、伊塔、考宇晨阳、郝坤、温家强、张涵毓、张允恒、雷贵艳、王振宁、赵梓雄等同学拍摄,由唐龙骞老师、郭今心老师编辑制作而成,在此对各位老师和同学表示衷心感谢!

本书的出版得到了山东大学化学与化工学院的大力支持,在此一并表示感谢!限于编者水平,书中难免存在不当之处,敬请读者批评指正。

编者

2022 年 1 月

目　录

Contents

第1章 有机化学实验的基本知识

1.1 有机化学实验的基本要求

有机化学是一门以实验为基础的学科，有机化学理论建立在实验的基础上，有机化学实验课的教学目的是通过实验使学生掌握有机化学实验的基本操作技能，验证和加深对有机化学基础知识和基本理论的理解，培养学生观察、记录和处理实验结果的能力，养成良好的实验工作习惯，同时也培养学生的批判性思维、创新思维、创新能力以及实事求是的科学态度和严谨细致的科学作风，为接受未来的挑战做准备，也为后续的学习和工作打下坚实的基础。

实验安全知识视频

为了保证实验的顺利进行和实验者的安全，在进入实验室前，都必须遵守以下相关规定。

1. 实验前认真预习实验内容，明确本次实验的目的、要求和内容，了解实验的基本原理、方法和步骤，了解所用试剂的危害及安全操作方法，学习相关实验仪器和设备的操作规程，提前写好预习报告。

2. 进入实验室时应穿实验制服，严禁穿拖鞋、高跟鞋或凉鞋、短裤、背心以及裙子入内，禁止披散长发。

3. 熟悉实验室工作的环境和安全规则，学会正确使用水、电、燃气、洗眼器、安全淋浴、灭火器、灭火毯和急救箱等。根据实验情况采取必要安全措施，如佩戴护目镜、面罩、手套等。如有受伤或事故发生，及时报告老师，并按照老师的指导进行处理。

4. 实验开始前认真检查仪器是否完整无损，是否齐全，使用时应小心谨慎，损坏以后及时报损。

5. 爱护公用仪器。公用仪器、工具及药品用后立即归还原处。使用药品前仔细阅读药品标签，厉行节约，药品按需取用，不可浪费，也不可将多取的药品倒回原试剂瓶。药品取用后，及时盖上瓶塞，避免瓶塞混用污染药品。不要随意改变实验室常用试剂和常用仪器的位置，如天平、灭火器、蒸馏水桶等。

6. 未经老师同意，实验时不得重做或更改实验方案。向老师报告所有异常情况，以减少操作危险。遵从教师指导，做到认真操作、仔细观察、积极思考和如实记录。

7. 严禁在实验室进食、吸烟、玩手机。实验过程中应集中注意力，不得做与实验无关的事情。正在进行的实验必须有人看守，不得擅自离开实验室。

8. 实验过程中要做到整洁有序，保持水槽、桌面、地面和仪器的洁净，火柴梗、废纸等废弃物不可随意乱扔，更不得丢入水槽中，以免堵塞下水道。

9. 实验室里，各种固体、液体废弃物应该严格按照要求分类回收处理，如有机溶剂、废水、固体废物和破碎玻璃等。

10. 轮流值日，离开实验室之前，仔细检查水、电、燃气阀及门窗是否关好，并把手清洗干净。

总之，要养成良好的实验习惯。

1.2 有机化学实验事故的预防和处理

化学实验首先要确保的是实验安全。有机化学实验所用的药品很多是有毒，易燃，具有腐蚀性、刺激性甚至爆炸性的物质，而化学反应又常在不同的条件下进行，需用各种热源、电器、玻璃仪器或其他设备，稍有不慎，便会造成火灾、爆炸、触电、割伤、烧伤或中毒等事故。因此，实验室的工作人员需要具有高度的安全防范意识，操作时要提高警惕，严格遵守操作规程，以避免事故的发生。为了预防和处理危险事故，应熟悉有关实验室安全的基本知识。

一旦有意外发生，应立即报告老师。牢记严重事故是不可逆转的，每个人的生命只有一次。

1.2.1 火灾的预防和处理

1. 预防

(1)使用易燃溶剂（如苯、丙酮、石油醚、二硫化碳或酒精等）时应远离火源。特别注意：操作乙醚时，严禁室内有明火。总之，要密切注意引燃源——明火、火花和热表面。

(2)蒸馏易燃液体时，要防止漏气，并注意通风，余气出口远离火源，最好用橡皮管通往室外。

(3)回流或蒸馏易燃液体时，不要用直火加热，应用水浴、油浴或加热套加热，同时应保持冷凝水从下端至上端流经的冷凝管畅通。

(4)不能用烧杯或敞口容器盛放易燃液体，且不能用明火加热。

(5)回流或蒸馏时，必须在蒸馏瓶内放入沸石，防止暴沸使液体溢出。若加热后发现未放沸石，则应停止加热，冷却后再放沸石。

(6)易挥发的可燃性废液不能倒入废物缸内，应倒入回收瓶中，并贴上标签，放入通风橱内，集中回收处理。

2. 处理

一旦发生火灾，应沉着冷静，切勿惊慌失措，及时采取措施。首先，应先切断电源和燃气，移去易燃易爆物品，然后采取适当的灭火措施。若火势不易控制，及时拨打 119 火警电话。

(1)在烧杯、蒸发皿或其他容器中的液体着火时，小火可用玻璃板、瓷板、石棉板或木板覆盖，将火熄灭。

(2)如果燃着的液体洒在地板或桌面上，应用干燥细沙扑灭。若着火液体是比水轻的有机溶液(如苯、石油醚等)，切勿用水扑救，因为这些液体会在水面上蔓延开来，反而使燃烧面积扩大。

(3)衣物着火时，尽快脱掉衣服或就地躺倒翻滚，也可用灭火毯等把着火部位包起来，就地滚动将火压灭。另外，也可以使用附近的安全淋浴将火浇灭。切勿奔跑或站立，以免火焰烧到头部。

(4)面对较大的火势，在确保自身安全的前提下，使用灭火器灭火。二氧化碳灭火器是有机实验室最常用的灭火器，内部装有压缩的液态二氧化碳，使用时二氧化碳气体喷出，用于小范围的油类、电器设备着火。使用二氧化碳灭火器时，一只手提灭火器，另一只手握在喇叭筒的把手上(不能手握喇叭筒，以防冻伤)。注意，不同类型的火灾需要不同类型的灭火器，不要用水扑灭化学火灾。不管采用哪一种灭火器，都要从火的外围向中心扑灭。

(5)扑灭燃烧的钠和钾时，千万不要用水，也不要使用四氯化碳灭火器。因为钠和钾会与水或四氯化碳发生剧烈反应，通常用干燥的细沙覆盖使其熄灭。

常用灭火器的性能及特点如表 1.2.1 所示。

表 1.2.1 常用灭火器的性能及特点

灭火器类型	成分	适用范围及特点
二氧化碳灭火器	液态二氧化碳	用于扑灭电器设备、小范围的油类及忌水的化学物品着火
水基型泡沫灭火器	水成膜泡沫灭火剂和氮气	用于扑灭石油及石油产品等非水溶性物质着火
干粉灭火器	主要成分是碳酸氢钠等盐类物质以及适量的润滑剂和防潮剂	用于扑灭油类、可燃气体、电器设备、精密仪器、图书文件及遇水易燃物品的初起火灾
洁净气体灭火器	七氟丙烷、三氟甲烷、IG541(氩气、氮气、二氧化碳)	用于扑灭油类、有机溶剂、精密仪器、高压电器设备着火

1.2.2 爆炸的预防

为了防止爆炸事故的发生，应注意以下几点：

1. 首先，操作者必须佩戴保护镜。若已佩戴隐形眼镜，应该事先告诉老师，并在整个实验过程中佩戴安全眼镜。

2. 不能在密闭体系内进行加热或反应。

3. 进行常压蒸馏时，蒸馏装置一定要与大气相通，否则会因蒸馏系统内气压增大而发生爆炸。进行减压蒸馏时，不能用平底烧瓶、锥形瓶、薄壁玻璃仪器等不耐压容器，实验前要检查玻璃仪器是否有破损。无论是常压蒸馏还是减压蒸馏，切记勿将液体蒸干。

4. 对于易爆炸的固体，如乙炔金属盐、苦味酸、苦味酸金属盐、三硝基甲苯等，切勿撞击或重压，以免发生爆炸，其少量残渣不能乱丢，应放入回收瓶中，并贴上标签，集中回收处理。

5. 有些有机化合物遇氧化剂，如硝酸、高锰酸盐、过氧化物等，会发生猛烈的爆炸或燃烧；碱金属（如钠）、锌粉和铂催化剂，操作或存放时应格外小心。

6. 使用易燃有机溶剂（特别是低沸点易燃溶剂），要远离明火，切勿将易燃溶剂倒入废物缸中，更不能用敞口瓶盛放易燃溶剂。

7. 使用醚类化合物之前，要检验是否有过氧化物，若有可用硫酸亚铁处理。

8. 反应过于剧烈时，应当控制加料速度和反应温度，必要时采取冷却措施。

1.2.3 触电的预防

电器设备使用前应检查线路连接是否正常、是否漏电，如遇轻微电击，应立即切断电源，检修仪器。使用电器时，应注意保持手、衣服和四周干燥，避免人体与仪器的导电部分直接接触，若电器上有水，应擦干后再使用。实验结束后应及时切断电源。

1.2.4 割伤的预防和处理

1. 预防

有机化学实验中使用玻璃仪器的最基本原则是不能对玻璃仪器的任何部位过度施压，并要注意以下几点：

(1)将玻璃管插入软木塞或橡皮塞中时，应用布包住，握住玻璃管的手应离塞子近些，慢慢将玻璃管旋转插入，以防割伤，必要时可在玻璃管上涂一些甘油以助滑入。

(2)已有裂痕或裂口的玻璃仪器切勿使用。洗涤玻璃仪器时应小心谨慎，以免损坏玻璃仪器。

(3)试剂瓶、量筒和表面皿等仪器不能加热。热的玻璃仪器不能突然触及冷的表面或冷水，否则将造成玻璃仪器破裂，使人受伤。

(4)不要把碎玻璃放在垃圾桶里,要置于专门盛碎玻璃的容器内。

2. 处理

熟悉实验室急救箱的位置。一旦受伤,无论重轻,都应立即向老师报告。

皮肉割伤后应先用消毒的镊子把伤口的玻璃屑取出,用蒸馏水洗净伤口,涂上碘酒,然后用纱布药棉包扎。若割伤较大,流血不止,应按住血管,或在伤口上方 5～10 cm 处用绷带扎紧,并送医院救治。

1.2.5　灼伤的预防和处理

1. 预防

皮肤接触了高温、低温或腐蚀性物质后均有可能被灼伤。为了避免灼伤,在接触这些物质时,要戴好防护手套和眼镜,并注意以下几点:

(1)处理热的物体和具有腐蚀性的化学药品时应特别小心,勿使其与身体直接接触。

(2)盛有液体的试管加热或煮沸时,管口不得对着自己或别人。在加热或反应进行时,不得靠近试管管口或烧瓶口观察反应。

(3)切勿倾水入酸,特别是在稀释浓硫酸时,必须将浓硫酸分次注入水中,同时加以搅拌。浓酸或浓碱等溶液加热时,应佩戴护目镜。

2. 处理

如果是热灼伤,应用冷水冲洗 10～15 min。如果是大面积灼伤,应立即就医。对于冷灼伤,不要加热,应用大量温水处理患处,并寻求医疗救助。

(1)若灼伤比较严重(如皮肤变黑、烧伤面积较大等),应先用消毒纱布敷在灼伤处,或用冷水浸湿的衣服遮盖以减轻疼痛,然后立即送医院救治,切勿在伤口上涂抹油或油性软膏。

(2)轻微的灼伤可涂苦味酸药膏或含有鞣酸的凡士林。

(3)被酸性物质灼伤时,先用大量水冲洗,然后用 1%～3%的碳酸氢钠溶液冲洗,再用水冲洗。

(4)被碱性物质灼伤时,先用大量水冲洗,然后用 1%的醋酸或 3%的硼酸溶液冲洗,然后用水冲洗。

(5)溴或苯酚滴在皮肤上时,可先用酒精洗去,然后在患处涂上甘油。

(6)若酸性液体溅入眼睛,应立即用水冲洗,然后用 1%碳酸氢钠溶液洗涤,最后用水冲洗。

(7)若碱性液体溅入眼睛,应立即用水冲洗,然后用 1%的硼酸溶液或 0.5%醋酸溶液洗涤,最后用水冲洗。

1.2.6　中毒的预防和处理

许多化学药品都有特殊的毒性,化学实验者必须了解化学物质的毒性。中毒主要是由吸入有毒气体或误服有毒物质所引起的，但也有从割伤或烧伤的皮肤处渗入人体的情况存在。为了防止中毒，应注意以下几点：

1. 有毒药品应妥善保管，不能乱放，其残渣也不能乱丢。

2. 勿让有毒药品触及五官或伤口，应穿戴防护服和手套，避免皮肤接触有毒药品。对产生刺激性或有毒气体的实验，应在通风橱中进行，或采用气体吸收装置。

3. 实验完毕后立即洗手。

实验过程中，若出现头晕、头痛等中毒症状，应立即离开现场，转移到通风良好的地方，并及时送往医院就诊。确定毒物的性质，对诊断治疗非常有帮助(最好能获取毒物样品、气体以及呕吐物等)。

1.3 有机化学实验废物的处理

有机化学实验室会产生各种固体、液体等废物，为保护环境，减少对环境的危害，废物处理可以按以下方式进行：

1. 实验室的废物不能随意丢弃，应按固体、液体，有害、无害等分类收集于不同的容器中，对于一些难处理的有害废物可送环保部门进行专门处理。

2. 破损的玻璃仪器和易燃物(如废纸或擦过易燃液体的织物)要分开放在不同的有盖子的废物箱中。手套、口罩、注射器等放置在专门的废物箱中按照医废处理。

3. 无害的固体废物(如软木塞、硅胶、硫酸镁等)可放入普通的废物箱。有毒的固体废物要用塑料袋封好，放在有标签的容器中。

4. 废弃的有机溶剂应装在有标签的容器中，并存放在通风处。

5. 卤代有机溶剂(如二氯甲烷、氯仿等)或含有卤化物的溶剂要用专门的容器回收，以备特殊处理。

6. 废弃的水溶液与有机废物要分开收集，以防它们之间发生剧烈反应，且它们储存和处理的方式不同。

7. 对于可能致癌的物质，处理时应格外小心，避免与身体接触。

任何情况下，未经处理的废弃物和有机溶剂都不能倒入水槽。

【知识拓展】

绿色化学的十二条原则

1.4 有机化学实验常用玻璃仪器和实验装置

1.4.1 常用玻璃仪器

1. 普通玻璃仪器

有机实验室常用的普通玻璃仪器如图 1.4.1 所示。

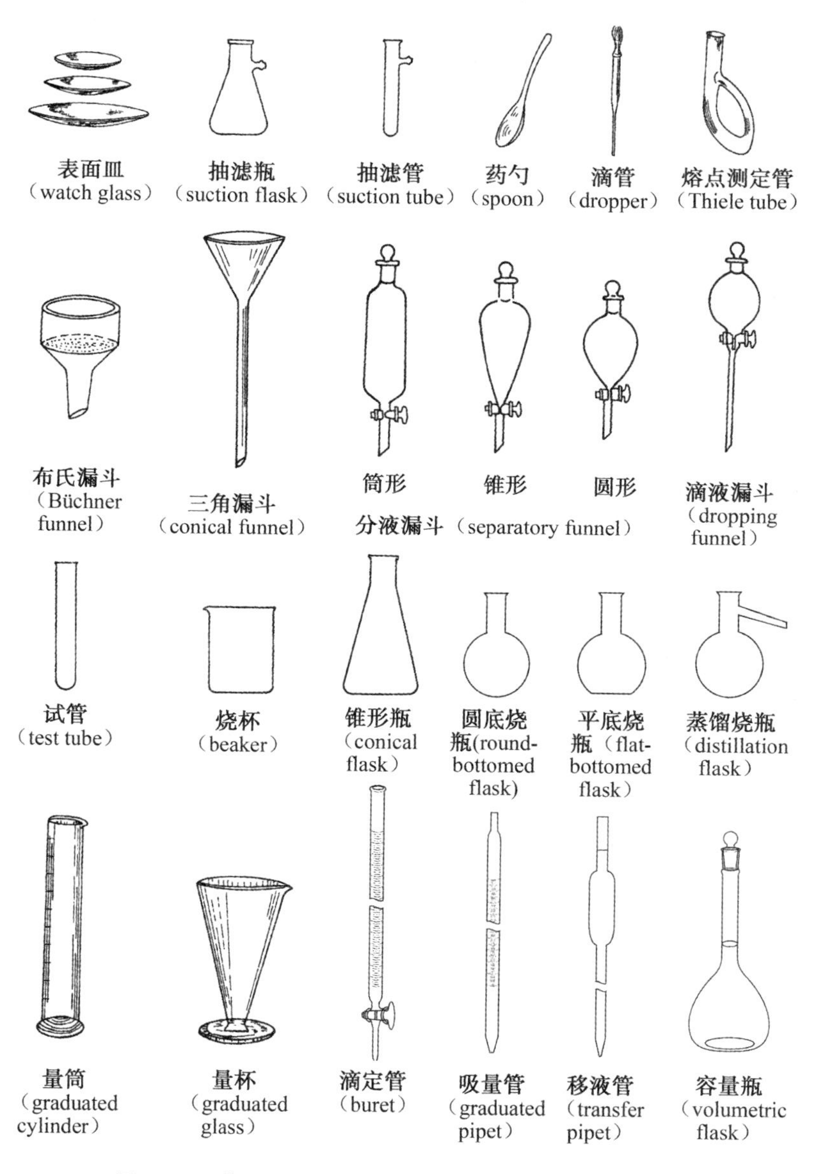

图 1.4.1 普通玻璃仪器(Non-standard taper glassware)

2. 标准磨口玻璃仪器

常用的标准磨口玻璃仪器如图 1.4.2 所示。

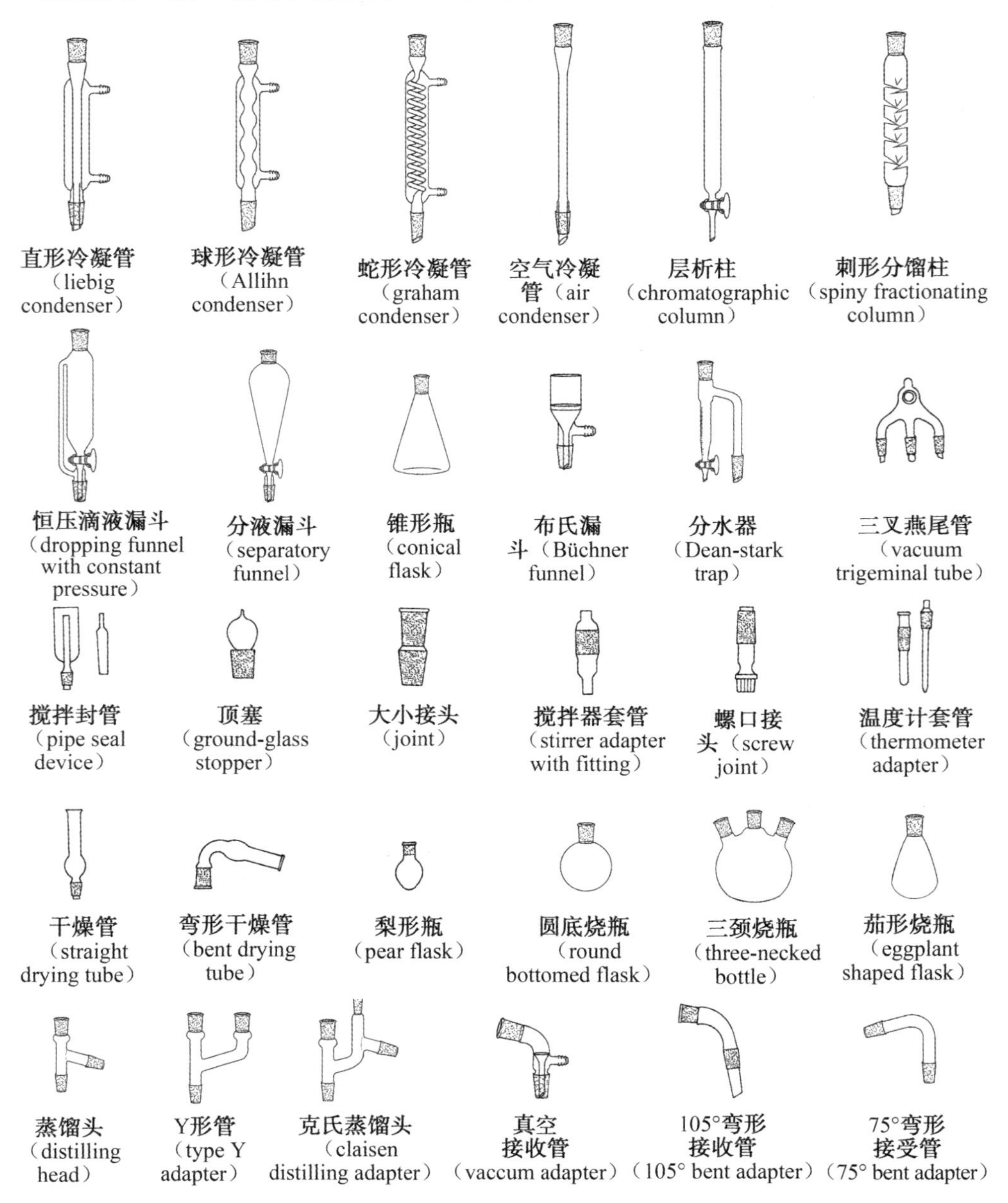

图 1.4.2 常用的标准磨口玻璃仪器(Common-used standard-taper ground glassware)

磨口玻璃仪器具有标准化、通用化、系列化等特点，使用方便。通常，标准磨口玻璃仪器有 10、14、19、24 等多种型号，这些数字编号指磨口玻璃仪器最大端直径的毫米数。另外，常用两个数字表示磨口玻璃仪器的大小，如 19/22，第一个数字表示磨口玻璃仪器最大端的直径，第二个数字表示磨口玻璃仪器的长度。

1.4.2 常用实验装置

常用的有机化学实验装置有回流装置、蒸馏装置、分馏装置、气体吸收装置、搅拌装置、过滤装置等。

1. 回流装置

进行有机反应时，需要对反应物进行长时间加热来提高反应速率，需要采用回流装置进行实验，常用的回流装置如图 1.4.3 所示。若反应需要防潮，可在球形冷凝管上端连接干燥管，如图 1.4.3(b)所示。若回流时有 HCl、SO_2 等水溶性气体产生，可在回流装置的出口连接气体吸收装置，如图 1.4.3(c)所示。图 1.4.3(d)(f)(g)所示装置在回流时可同时滴加液体。图 1.4.3(f)为索氏提取器，将待测样包在滤纸中放入提取管内，通过回流浸取样品中的脂类物质。

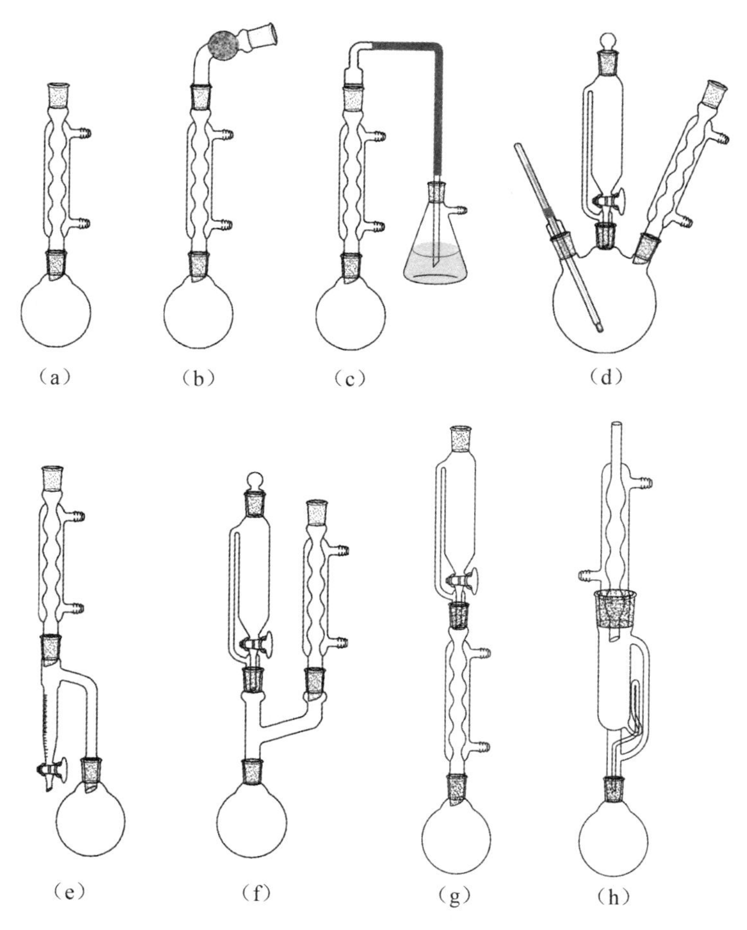

图 1.4.3 常用的回流装置(Common reflux apparatus)

2. 蒸馏装置

常压蒸馏装置和减压蒸馏装置如图 1.4.4 所示。蒸馏常用于分离两种以上沸点相差

25 ℃以上的液体，并可用于除去有机溶剂。图 1.4.4(a)所示装置为常压蒸馏装置，常用水冷凝管将冷凝水从下端回流至上端。若蒸馏沸点在 140 ℃以上的液体，可将直形水冷凝管换成空气冷凝管，避免蒸气温度过高使冷凝管炸裂。图 1.4.4(b)所示装置为减压蒸馏装置，常用于易分解、氧化、聚合等物质的蒸馏。如果在蒸馏过程中需补加溶剂，可在蒸馏头上端加装滴液漏斗。

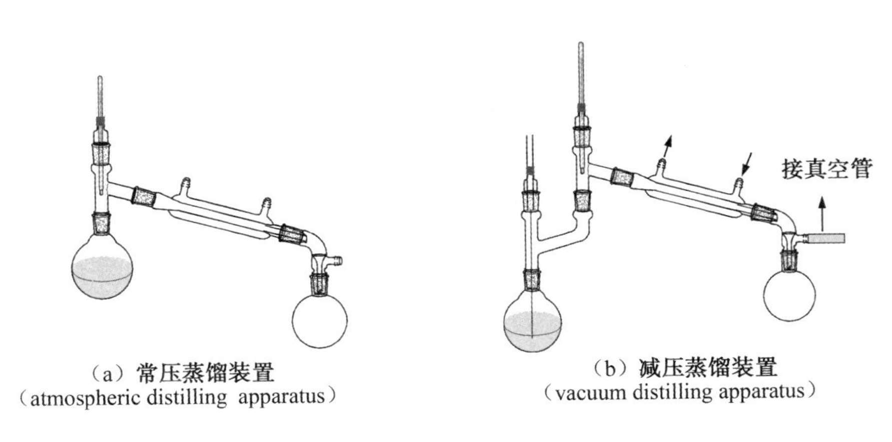

(a) **常压蒸馏装置**
(atmospheric distilling apparatus)

(b) **减压蒸馏装置**
(vacuum distilling apparatus)

图 1.4.4　蒸馏装置(Distillation apparatus)

3. 分馏装置

图 1.4.5(a)(b)(c)所示装置为球形分馏柱、韦氏分馏柱(刺形分馏柱)和填充式分馏柱，图1.4.5(d)为常见的分馏装置。当混合试剂中两种液体组分沸点相差较小时，蒸馏无法分离，需要通过分流的方法进行精确分离。

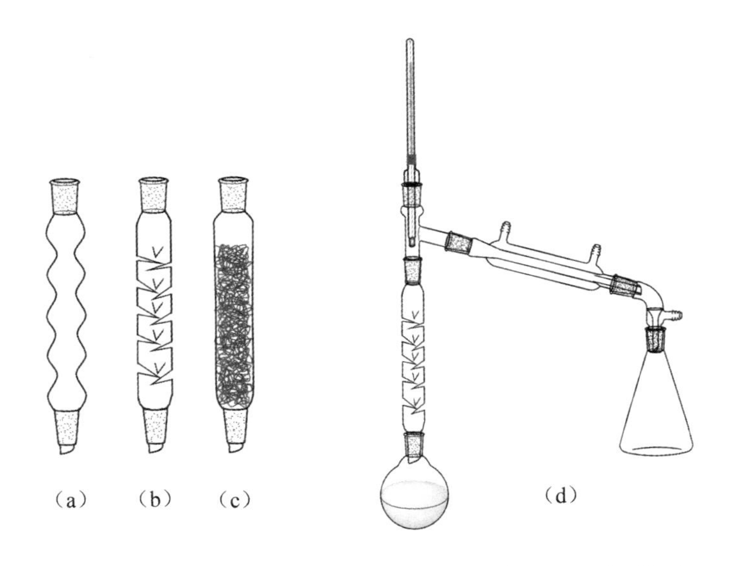

图 1.4.5　分馏柱和分馏装置(Fractionating column and fractional distillation apparatus)

4. 气体吸收装置

气体吸收装置用于吸收反应中产生的可溶性气体，如 HCl、SO_2 等。常用的气体吸收装置如图 1.4.6 所示，图 1.4.6(a)和图 1.4.6(b)所示装置可用于吸收少量气体。图 1.4.6(a)中漏斗倾斜，一半没在水中，防止气体外溢；一半在水面之上，防止水倒吸至反应瓶中。图 1.4.6(c)所示装置可用于有大量气体产生或气体产生速度较快的反应。水从上端流入抽滤瓶，在恒定液面溢出。玻璃管口在水面之下，防止气体进入大气。

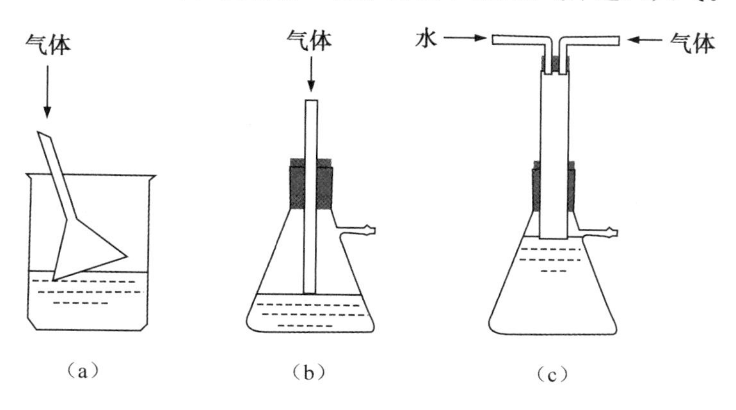

图 1.4.6　常用的气体吸收装置(Gas absorption apparatus)

5. 搅拌装置

搅拌装置能够使非均相反应迅速均匀混合，防止局部过浓或局部过热现象出现，避免副反应发生，常见的搅拌装置如图 1.4.7 所示。图 1.4.7(a)所示装置为带搅拌可测温的回流装置，图 1.4.7(b)所示的装置可以在搅拌过程中滴加液体，图 1.4.7(c)所示的装置可以同时实现搅拌、测温、回流和滴加液体。

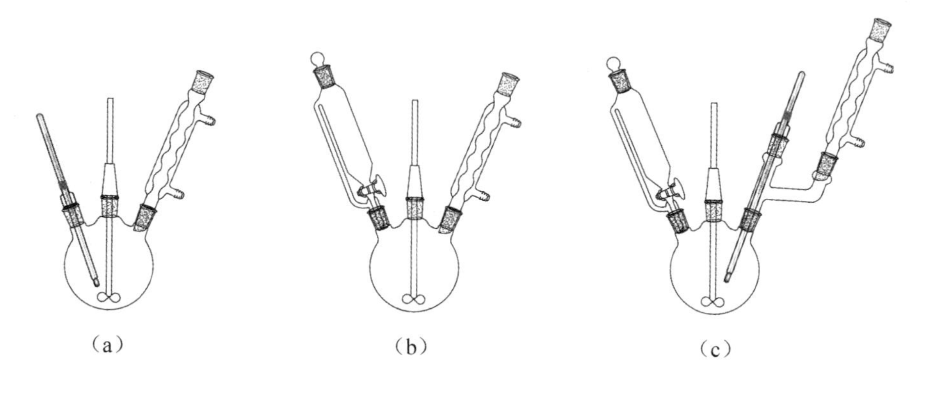

图 1.4.7　搅拌装置(Stirring apparatus)

1.4.3 使用玻璃仪器的注意事项

使用玻璃仪器时需要注意以下几点：

1. 使用玻璃仪器时要轻拿轻放，以免弄碎。

2. 除烧杯和试管外，其他玻璃仪器均不能用火直接加热。

3. 锥形瓶、平底烧瓶不耐压，不能用于减压操作。

4. 带活塞的玻璃器皿用后应及时清洗，并在活塞与磨口之间垫上纸片，以防粘连。

5. 如果接头粘住了，可以用木棍轻轻敲打使其松动。如果这个方法不行，可以在热水或蒸汽浴中加热接头，也可以用吹风机来加热接头。

6. 分液漏斗的塞子是配套磨口的，如果不是原装的，可能会漏液，所以要注意保护。

7. 温度计的水银球处玻璃很薄、易碎，使用时应小心。不能将温度计当作搅拌棒使用。温度计使用后应先自然冷却再冲洗，测量范围不得超出温度计刻度范围。

8. 水银温度计若不慎打碎，应尽可能将水银珠收集起来，并用硫黄粉覆盖水银珠和撒落水银的地面，破碎的温度计应插入盛有硫黄的有盖容器中。

1.5 有机化学实验玻璃仪器的清洗与干燥

1.5.1 玻璃仪器的洗涤

实验所用的玻璃仪器的洁净程度，直接影响着实验结果。因此，实验人员要学会洗涤及干燥的方法，养成每次实验完成后立即洗净仪器的习惯。

清洗玻璃仪器的一般方法是将玻璃仪器和毛刷淋湿，用毛刷蘸取去污粉或洗涤剂清洗玻璃器皿内外壁，除去污物后再用自来水冲洗。当洁净度要求较高时，可依次用洗涤剂、自来水、蒸馏水（或去离子水）清洗。仪器倒置，器壁不挂水珠时，即已洗净。

严禁盲目使用各种化学试剂和有机溶剂来清洗仪器，这不仅会造成浪费，而且还可能带来危险，污染环境。

玻璃仪器也可以用超声波清洗器来洗涤，仪器放在一定浓度的洗涤剂溶液中，利用声波的震动和能量洗涤，方便快捷。

1.5.2 仪器的干燥

仪器的干燥与否有时甚至是实验成败的关键。一般实验将洗净的仪器倒置一段时间后，若没有水迹，即可使用。有些实验要求无水操作，这时可将仪器放在烘箱中烘干。但要注意在用烘箱干燥仪器时必须等烘箱冷却到室温后才可以取出仪器，否则由于仪器温度较高，遇到冷空气时会有水汽在器壁冷凝下来，而达不到干燥的目的。

若需快速干燥，可用少量有机溶剂（如丙酮或95%乙醇）淋洗玻璃仪器，然后再用电吹风吹干即可。用这种方法干燥仪器时，要让玻璃仪器里的水完全流尽，用少量(约10 mL)有机溶剂淋洗玻璃仪器，用过的溶剂应倒回回收容器中，然后用电吹风吹干玻璃仪器。

1.6 加热与冷却

1.6.1 加热

加热是一项重要的实验室技术，它可以增加化学反应的速度，在重结晶和提纯过程中促进固体的溶解，也用于液体的蒸馏等。实验室常用的热源有煤气灯、酒精灯、电热套和电炉等。从加热方式来看，有直接加热和间接加热两种方法。在有机化学实验中一般不用直接加热；为了保证加热均匀，常常使用间接加热的方法。间接加热传热介质有空气、水、有机液体及石沙等。常用的加热方法有以下几种：

1. 空气浴

空气浴就是利用空气间接加热，对于沸点高于 80 ℃的液体均可采用空气浴，如石棉网上加热(这是最简单的空气浴)。但空气浴会导致受热不均匀，不能用于回流低沸点易燃液体、减压蒸馏等。

最常用的空气浴加热仪器是电热套，它可加热到 300 ℃，常用于回流加热。但用电热套加热时，温度不易控制，不适用于蒸馏或减压蒸馏。安装电热套时，要使反应瓶外壁与电热套内壁保持 1～2 cm 的距离，以利于空气传热，并防止局部过热。注意：不要把反应液撒到加热套上。

2. 水浴

水浴是最常用的热浴方法，加热温度不超过 100 ℃的操作，均可采用水浴加热。但含有钾、钠的操作切勿用水浴加热。水浴加热时，水面一定要高于反应液的液面。

3. 油浴

加热温度在 100～250 ℃的实验可用油浴。油浴的优点是温度易控制，容器内的反应物受热均匀。容器内反应物的温度一般比油浴温度低 20 ℃左右。常用的浴液有以下几种：

(1)甘油：可以加热到 140～150 ℃，温度过高会分解，并释放出难闻的气味。甘油吸水性强，长时间放置未使用的甘油在使用前应加热去除所吸收的水分。

(2)植物油：如菜籽油、蓖麻油、花生油等，可以加热到 220 ℃，常常还要加入 1%对-苯二酚作抗氧化剂。植物油温度过高会分解，达到闪点会燃烧，使用时应小心。

(3)液态石蜡：可加热到 200 ℃左右，高温易燃。

(4)硅油：可加热到 250 ℃，透明度好，是理想的浴液，但价格较贵。

用油浴加热时，油量不能过多，以超过反应液面为宜，否则受热后，容易溢出发生火灾。热油中也不能溅入水，否则加热时会产生泡沫或引起爆溅。加热完毕，取出反应器时，应用铁夹夹住反应器离开液面悬置片刻，待反应器壁上油滴完后，用纸或干布擦净器壁。加热时要注意安全，防止意外起火。

4. 沙浴

沙浴一般是将沙子装入干燥的铁盘中，将反应容器半埋入沙中加热。沸点高于80 ℃的液体均可采用沙浴，沙浴特别适于要求加热温度220 ℃以上的实验。但沙浴传热慢，温度上升慢，不易控制。

5. 微波加热

在20世纪90年代，用于实验室的微波反应器开始商业化，现在微波加热或微波辅助有机合成(MAOS)在许多实验室比较常见。与传统的加热方法相比，现代微波设备具有安全、反应速度快、节能、产率高和副产物少等优点。对于微波加热有机物质，最重要的机理是偶极极化和离子传导。实验室微波反应器有多模式和单模式两种。多模式反应器外观类似厨房微波炉。

注意：家用微波炉绝对不能用于化学反应加热。这些设备不能监控反应温度和压力，可能导致火灾和爆炸。只有科研专用微波反应器才可在实验室中使用，并严格按照要求操作。

1.6.2 冷却

在有机化学实验中，有时需要进行低温冷却操作，如重氮化反应一般在0～5 ℃进行。有些有机合成反应会产生大量的热，使得反应温度迅速升高，如果控制不当，可能引起副反应或使反应物蒸发，或者逸出反应器，甚至引起爆炸。为了将温度控制在一定范围内，也要进行适当的冷却。有时为了降低溶质在溶剂中的溶解度或加速结晶析出，也要采用冷却的方法。常用的冷却方法有以下几种：

1. 冰-水浴

水是常用的低成本、高热容的冷却剂。冰-水浴常用水和碎冰的混合物作冷却剂，可冷却至0～5 ℃。由于和容器的接触面积大，冰-水浴的冷却效果比单用冰块好。如果水不影响反应进行，也可把碎冰直接投入反应器中，以便更有效地保持低温。

2. 冰-盐浴

反应要在0 ℃以下进行操作时，常用按不同比例混合的碎冰和无机盐作为冷却剂。可把盐研细，和碎冰按一定比例混合。冰-盐浴的混合比例及能达到的最低温度如表1.6.1所示。

表 1.6.1 冰-盐浴的混合比例能达到的最低温度

无机盐	质量配比(盐∶冰)	最低温度/℃
NH_4Cl	1∶4	−15
NaCl	1∶3.3	−21
$NaNO_3$	1∶2	−18
$CaCl_2\cdot 6H_2O$	1∶1	−29
$CaCl_2\cdot 6H_2O$	1∶0.7	−55

3. 干冰或干冰与有机溶剂混合冷却

干冰(固体的二氧化碳)可冷却至−60 ℃以下。若将干冰与乙醇、异丙醇、丙酮等混合,可冷却到−78～−50 ℃,但要注意混合时会猛烈起泡。

4. 液氮

液氮可以冷却至−195.8 ℃。液氮和干冰这两种冷却剂应放在杜瓦瓶(广口保温瓶)中或其他绝热效果好的容器中,以保持其冷却效果。使用杜瓦瓶时,要佩戴护目镜。

5. 低温浴槽

低温浴槽是一个小冰箱,冰室口向上,工作槽内胆及机箱外壳全部采用不锈钢材料,内装酒精,设有循环泵,可把槽内被恒温液体外引到冷凝器等。低温浴槽可以装温度计等指示器,反应瓶浸在酒精液体中,适于−30～30 ℃范围的反应。

注意:温度低于−38 ℃时,水银会凝固,因此不能用水银温度计。对于较低的温度,应采用添加了少许颜料的有机溶剂(乙醇、甲苯、正戊烷)的低温温度计。

1.7 干燥与干燥剂的使用

干燥剂与干燥剂的使用

1.8 实验预习、实验记录及实验报告

1.8.1 实验预习

每位实验者都应准备一本规范化的实验记录本。在每次实验前必须认真预习，作好充分准备。预习报告包括以下内容：

1. 实验的目的、要求。
2. 实验原理，包括反应式（主要反应、主要副反应）、反应机理等。
3. 主要试剂和产物的物理常数（查阅手册或辞典）及主要试剂的用量（克、毫升、摩尔）和规格（纯度）以及实验仪器。
4. 反应主要装置图、实验的简要步骤及操作原理。
5. 粗产物纯化的流程图。
6. 实验中可能出现问题的防范措施和解决办法。

1.8.2 实验记录

实验者必须养成一边实验一边做记录的习惯。记录内容包括实验的全过程，如加入样品的数量、操作过程所观察到的现象（包括温度、颜色的变化）和时间等。记录要求实事求是，文字力求简明扼要，特别是当观察到的现象和预期相反或与实验教材所叙述的内容不一样时，应记录下实验的真实情况，并用明显的标记注明，以便探讨其原因。其他各项，如在实验过程中的一些准备工作、现象解释、称量数据等，可以记在备注栏内，实验记录的格式如下：

1. 实验名称：
2. 时间：
3. 气温和空气湿度：
4. 操作步骤：
5. 现象：
6. 备注：

实验结果可整理成如下格式：

1. 产物名称：
2. 产品的物理性状：
3. 产量：______g　　产率：______%

实验完毕后，须将实验记录本交给指导老师签字，方可离开实验室。

1.8.3 实验报告

实验报告是实验结束后对实验过程的情况总结、归纳和整理，对实验现象和结果的分析讨论，也是实验的重要组成部分。实验报告要求真实可靠，数据完整，文字简练，条理清楚，书写工整。实验报告包括以下几部分：

1. 实验名称和时间。
2. 实验目的和要求。
3. 实验原理(如反应式、主要副反应等)。
4. 仪器与主要试剂(用量、规格)。
5. 主要装置图。
6. 实验步骤及现象记录。
7. 实验结果(如 R_f 值、熔点、产量、产率计算等)。
8. 讨论。

讨论是实验报告中最重要的部分，需要分析和解释实验结果，对操作的经验教训和实验中存在的问题提出改进建议。例如，可讨论本次实验是否达到了预期实验目标，如果没有达到目标，该如何改进；讨论实验中存在的失误和原因，以及整个反应结果的影响等。

【思考题】

1. 请列出所在实验室的安全防护措施。
2. 找到实验室的急救箱，检查里面的医疗用品。若实验过程中眼睛不慎被碱灼伤，应该怎么处理？

【知识拓展】

有机化学实验的新技术与新方法

1.9 有机化学的文献资料

有机化学的文献资料

第2章　基础有机化学实验

2.1　熔点的测定

2.1.1　熔点的测定方法

熔点的测定
（操作视频）

【实验目的】

1. 明确熔点测定的原理和意义。

2. 掌握熔点测定的方法。

【实验原理】

晶体的熔点是在标准大气压下，固、液两态平衡时的温度。熔点是晶体物质的重要物理常数。纯粹的固体有机化合物一般都有固定的熔点，并且熔距或熔程（即开始熔化到完全熔化的温度间距）很短，只有0.5～1 ℃。

图2.1.1为相变的温度-时间变化图。从图中可以看出，固体受热时，从固体到液体的变化表现在熔点处的曲线斜率突然变化。达到熔点时，固体晶格的分解需要吸收热量。如果固、液两态之间保持平衡，温度就不会上升，直到所有的固体熔化。在实际的熔点测定中，固体完全熔化时的温度稍微有所上升。因此，观察到的熔点通常为至少1 ℃的范围。当有杂质存在时，固体有机化合物的熔点降低，熔距增大。

熔点是晶体物质的重要物理常数。熔点的测定常被用来鉴定有机化合物，或判断它的纯度。

测定熔点的方法有毛细管法和显微熔点测定法，这里主要介绍毛细管法。显微熔点测定法需要使用WRR熔点测定仪，该仪器的使用方法将在后面介绍。

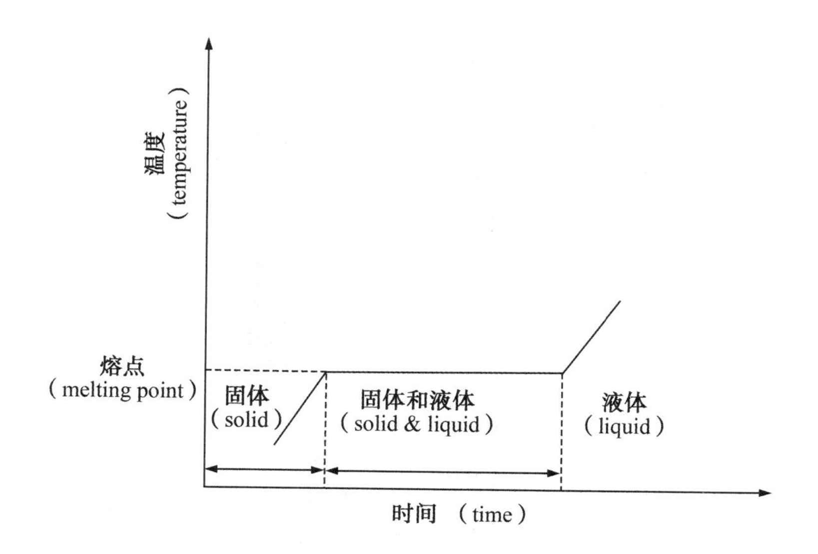

2.1.1 相变的温度-时间变化图(Phase change with time and heat)

【仪器与试剂】

仪器:熔点测定管,毛细管若干,温度计(150 ℃),酒精灯,表面皿,长玻璃管(1 cm×50 cm),小橡皮圈,铁架台(带铁夹)。

试剂:液体石蜡,尿素,苯甲酸,尿素和苯甲酸的混合物,未知物化合物 A、B。

【实验步骤】

1. 毛细管的制备:取市售毛细管数根,将其一端封口。封口时将一端放在酒精火焰上,慢慢转动加热,转速必须均匀(封口处不能弯曲,不能鼓成小球)。

2. 样品的填装:取少许样品于表面皿上,用玻璃棒研成细粉,聚成小堆。将毛细管开口的一端插入其中,样品将被挤压入毛细管中。然后将毛细管开口端朝上,从一根 30~50 cm 的玻璃管中滑落至实验台上,重复几次,使样品紧密平整地填装在毛细管底部,所装样品高度为 2~3 mm。

3. 仪器装置:向熔点管中倒入浴液,浴液面要高出熔点管的侧管上口 1 cm。然后将熔点管夹于铁架台上,用橡皮圈将装好样品的毛细管上端套在温度计上,橡皮圈不要接触浴液,毛细管下端装有样品的部分应紧靠在水银球的中部。最后通过一个开口的软木塞将温度计插入熔点管中,水银球的位置恰好在侧管上口与下口之间。仪器装置如图2.1.2 所示。

4. 熔点的测定:测定已知物的熔点时,可先查阅化学手册,获知真实熔点。实验中用酒精灯加热熔点管的弯曲侧管底部,使温度按每分钟 5~6 ℃的速度上升,当低于预期的熔点 10~15 ℃时,改为小火加热,使温度每分钟上升约 1 ℃。

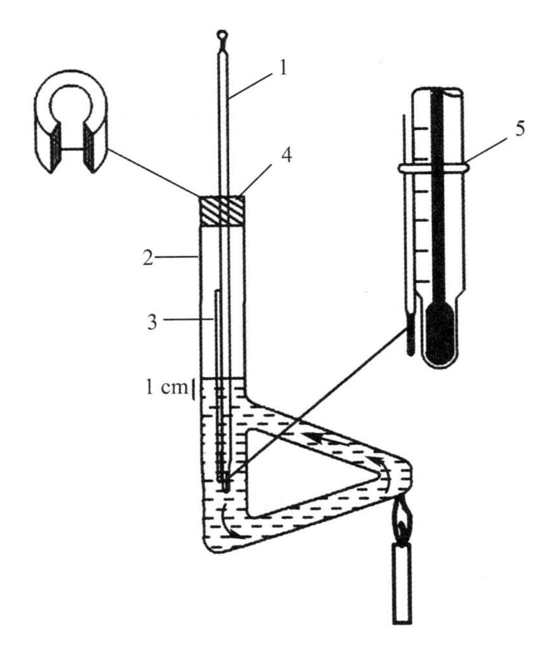

1—温度计(thermometer);2—熔点测定管(thiele tube);3—毛细管(capillary tube);4—缺口软木塞(cork notched to permit escape of air);5—橡胶圈(rubber ring)

图 2.1.2　熔点测定装置图 (Thiele apparatus)

仔细观察温度上升和毛细管中样品的情况。记下样品开始塌落、润湿并出现微小液滴时的温度(初熔)及固体全部消失时的温度(全熔),即为被测物质的熔点。如某物质在121 ℃时初熔,122 ℃时全熔,可记录熔点为121～122 ℃。

测定未知物的熔点时,应先将样品填装好两根毛细管,用一根迅速测得熔点的近似值,待浴液温度下降约30 ℃后,置换第二根毛细管,用小火加热,仔细地测定样品的精确熔点。

对于易升华的物质,应用两端封闭的毛细管,毛细管全部浸入浴液中。对于易分解的样品(在到达熔点时,可见其颜色变化,样品有膨胀和上升现象),可把浴液预热到距熔点20 ℃左右时,再插入样品毛细管,改用小火加热测定。若到一定温度时,样品完全分解而不熔化,应在熔点值中标记上“分解”。例如,某物质的熔点可记录为131.5 ℃(分解)。

【解释说明】

1. 毛细管填装样品的量直接影响样品的测定,样品过多,使其熔程变长;样品过少,不易观察。

2. 加热速度对温度计读数影响较大,加热过快,读数偏高;加热过慢,则读数偏低。

【操作注意事项】

1. 接近熔点时升温速度要降低,升温太快对所测熔点将产生影响。

2. 样品一定要研磨的极细才能使装样结实,这样受热才均匀。如果有空隙,不易传热,会影响熔点测定结果。

3. 浴液一般用浓硫酸或液态石蜡。浓硫酸可加热到 250~270 ℃,加热时必须小心,不可使温度过高,以免硫酸分解,放出 SO_3。此外,要防止硫酸触及皮肤而造成灼伤。液态石蜡可加热到 200~210 ℃,但其蒸气可燃,操作时应多加注意。

4. 温度计冷却后,用废纸揩净浓硫酸后,方可用水清洗,否则温度计易炸裂。

【废物处理】

实验完成后,废熔点管、橡皮圈、废纸倒入指定的废物桶内。

2.1.2 WRR 熔点测定仪的使用方法

WRR 熔点测定仪的使用方法

【思考题】

1. 比较两种测定方法,影响熔点测定准确性的因素有哪些?

2. 有两瓶白色结晶状的有机固体,分别测得熔点为 130~131 ℃和 130~130.5 ℃,如何鉴别两种物质是否相同?

2.2 沸点的测定及常压蒸馏

常压蒸馏
(实验原理)

沸点的测定
(操作视频)

【实验要求】

1. 了解普通蒸馏和沸点测定的原理和意义。

2. 掌握常量法测定沸点的操作方法。

【实验原理】

由于分子运动,液体受热后,其分子从液体表面逸出,并在液面上部形成蒸气压。随着温度升高,蒸气压增大,当蒸气压与外界大气压相等时,有大量气泡从液体内部逸出,液体开始沸腾,此时的温度就是该物质在此压力下的沸点。显然,沸点与外界大气压力有关。通常,沸点是指在标准大气压(101.3 kPa)下液体的沸腾温度。例如,水的沸点为 100 ℃,即在 101.3 kPa 气压下,水在 100 ℃时沸腾。在其他气压下的沸点应注明气压。例如,在 70 kPa 气压下,水在 90 ℃时沸腾,这时水的沸点可以表示为 90 ℃/ 70 kPa。

纯粹的液态有机化合物在一定大气压下具有恒定的沸点,且沸点的温度范围(称为沸程)较小,通常为 1~2 ℃。挥发性及非挥发性杂质都会影响沸点。

沸点是液态物质的重要物理常数。沸点的测定也常用于鉴定有机物质或判断其纯度。

值得注意的是,具有恒定沸点的液态物质不一定都是纯净的化合物,因为有些液态物质常与其他组分组成二元或三元共沸混合物,它们也有恒定的沸点。例如,氯化氢与水可形成沸点为 108.5 ℃的二元恒沸混合物(水的质量分数为 79.8%),乙醇与水可形成沸点为 78.2 ℃的二元恒沸混合物(水的质量分数为 4.4%)。

根据样品用量的不同,测定沸点的方法可以分为常量法和微量法两种。常量法测定沸点的基础是蒸馏。把液体加热变为蒸气,然后再使蒸气冷凝并收集于另一容器中的操作叫作蒸馏。蒸馏是分离或提纯有机物质常用的方法之一。微量法用微量沸点管和毛细管测定沸点。

【仪器与试剂】

仪器:磨口蒸馏烧瓶,直形冷凝管,真空尾接管,蒸馏头,标准接头(∅14 mm)、温度计(100 ℃),温度计接头(∅14 mm),量筒,浴锅,铁架台(附铁夹),橡皮管,沸石少许,橡皮管。

试剂:无水乙醇(A.R)。

【实验步骤】

1. 常压蒸馏装置

常压蒸馏装置如图 2.2.1 所示,主要包括蒸馏瓶、冷凝器及接收器三部分。

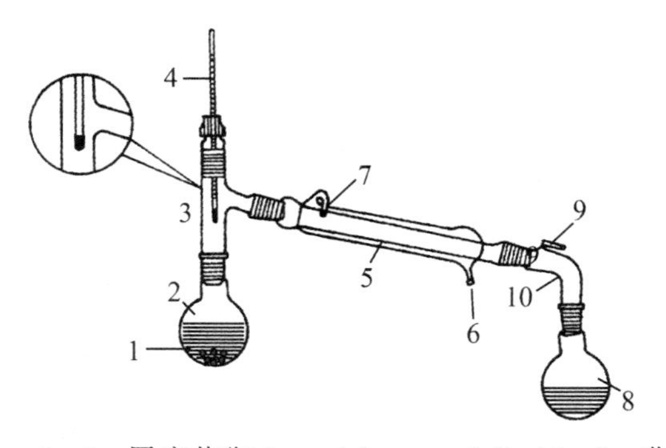

1—沸石(boiling stones);2—圆底烧瓶(round bottomed flask);3—蒸馏头(distillation head);4—温度计(thermometer);5—冷凝管(condenser);6—进水口(water in);7—出水口(water out);8—接收瓶(receiving flask);9—抽气支管(vacuum/gas inlet);10—真空接收管(vacuum adapter)

图 2.2.1 常压蒸馏装置(Simple distillation apparatus)

安装时,取一个磨口蒸馏烧瓶与蒸馏头相连,将温度计穿过橡皮管插入温度计接头中,再与蒸馏头相接。调整温度计的位置,蒸馏时水银球能完全被蒸气包围。通常水银球的上端应恰好位于蒸馏头支管的底边所在的水平线上(见图 2.2.1),把蒸馏瓶夹在铁架台上,并放入水浴中。

冷凝管夹在另一个铁架台上(夹住冷凝管的中上部),冷凝管的斜度应与蒸馏头侧管的斜度相同。侧管应与冷凝管相连。

冷凝管下端的侧管用橡皮管与水源连接，上端的侧管接橡皮管并通入水槽中，冷凝管的末端与真空尾接管相连，其末端与接收瓶（磨口蒸馏烧瓶，本实验采用 10 mL 量杯）相接。

2. 沸点的测定

用干燥的量筒量取 20 mL 无水乙醇，倒入蒸馏烧瓶中，加几粒沸石于瓶中以防止爆沸。按图 2.2.1 安装好蒸馏装置，整个装置要安装平稳，接口处不得漏气。在蒸馏过程中要随时检查接口是否严密以防蒸气逸出。将蒸馏瓶放入水浴中，水面略高于瓶内液体的液面。缓慢开启自来水管，使水流平稳流过冷凝管。加热水浴使水沸腾。注意观察蒸馏瓶内液体沸腾、蒸发的情况及温度的变化，当瓶内液体沸腾，蒸气前沿到达温度计时，温度计读数急剧上升，适当调节热源，使馏出液每秒滴 1～2 滴。当第一滴冷凝液滴入接收瓶时，记下温度。继续加热，直至蒸馏瓶中仅剩少量液体时（不要蒸干），再记录一次温度。起始及最终的温度就是样品的沸点，两个温度之差即为样品的沸程。根据沸程的大小，便可判断样品是否纯净。

蒸馏完毕，应先停火，移走热源，待稍冷却后，关好冷却水，拆除仪器。

3. 微量法测沸点

取一段直径 4～5 mm、长 5～6 cm 的玻璃管，密封其一端（或用已制好的小试管），内放试液 2～3 滴。在管内放入一根直径约 1 mm、上端密封的毛细管（长 7～8 cm），按照图 2.2.2 将微量沸点管安装好后，将其用橡皮圈固定在温度计的一侧，其底部样品的位置在水银球中间。然后将整套装置放入浴液中。

慢慢加热浴液，使温度均匀上升，当达到样品沸点时，可看到内管下端有连续的小气泡出现。当管内出现大量快速而连续的气泡时，说明毛细管内的空气已完全被试液的蒸气所换出，立即停止加热。随着温度的降低，气泡逐渐减少，记下最后一个气泡刚欲缩回毛细管中时的温度，也就是液体的蒸气压与大气压平衡时的温度，即试液的沸点。

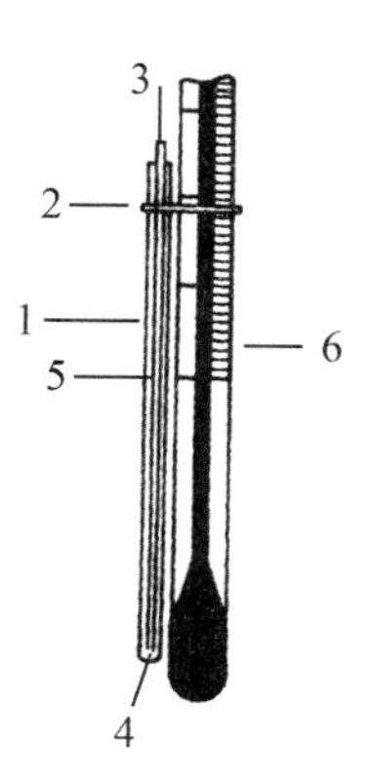

1—玻璃管（glass tube）；2—橡胶圈（rubber band）；3—毛细管封口端（closed end of capillary tube）；4—毛细管开口端（open end of capillary tube）；5—毛细管（capillary tube）；6—温度计（thermometer）

图 2.2.2 微量法测沸点（Microscale boiling point determination）

【结果记录】

实验完成后，将结果填在表 2.2.1 中。

表 2.2.1　结果记录表

流出量	第 1 滴	2 mL	4 mL	6 mL	8 mL	10 mL
沸点/℃						

【解释说明】

1. 选用蒸馏瓶时，一般应使所盛液体的量不少于蒸馏瓶容积的 1/3，不多于蒸馏瓶容积的 2/3。

2. 为了避免液体在加热过程中出现过热现象和液体暴沸现象，蒸馏前需加沸石。若中途停止蒸馏，在继续蒸馏前，应补充新的沸石。这是因为原有的沸石空隙内已充满了液体，已失去防暴沸作用。

3. 冷凝器的选择根据液体的沸点而定，液体沸点高于 130 ℃的应用空气冷凝管，低于 130 ℃的则用冷水冷凝管。

4. 热源的选择根据液体的沸点而定。沸点在 80 ℃以下的易燃液体宜用沸水浴，高沸点的液体可用油浴或沙浴。有机液体一般不用直火加热。

【操作注意事项】

1. 普通蒸馏装置不能密闭，以免瓶内蒸气压过大而引发爆炸。常压蒸馏时，真空尾接管应与大气相通。若蒸馏液具有毒性或腐蚀性，应由橡皮管导入特殊装置或吸收液中；若蒸馏液易吸水，则在真空尾接管的侧管上装一个干燥管，再与大气相通。

2. 温度计水银球的上沿与蒸馏烧瓶支管口的下沿在同一水平线上。

3. 火源与接收器之间应保持一定距离，以免引发火灾。

4. 加热前一定要先通冷凝水，冷凝水应“下进上出”。实验完毕，应先撤去火源，稍等几分钟，等温度稍微冷却后再停止通水。

5. 蒸馏速度不应太快或太慢。在蒸馏过程中，应始终保持温度计水银球上有一稳定的液滴，这是气液两相平衡的象征，这时温度计的读数便能代表液体的沸点。

【废物处理】

实验完成后，将收集的乙醇倒入专门的回收容器中进行回收处理。

【思考题】

1. 蒸馏装置中，温度计位置对温度读数有何影响？什么是最佳位置？

2. 蒸馏有机物时为什么不能用直火加热？

3. 蒸馏时，沸石的作用是什么？是否可以将沸石加入接近沸腾的溶液中？重新蒸馏或补加溶液后，是否需要重新加入沸石？

4. 微量法测沸点的原理是什么？

2.3　减压蒸馏

【实验目的】

1.了解减压蒸馏的基本原理及其应用。

2.掌握减压蒸馏仪器装置的安装及操作技术。

【实验原理】

减压蒸馏是分离和提纯有机化合物的一种重要方法，特别适用于常压条件下沸点较高或者常压蒸馏时未达沸点即已受热分解、氧化或聚合的物质。沸点大于 200 ℃的液体一般需用减压蒸馏分离提纯。

液体的沸点是指它的蒸气压等于外界大气压时的温度，所以液体沸腾的温度随外界气压的降低而降低。若用真空泵连接盛有液体的容器，可使液体表面上压力降低，即可降低液体的沸点。这种在较低压力下进行蒸馏的操作称为减压蒸馏。在进行减压蒸馏之前，应先查阅欲提纯的化合物在所选压力下的相应沸点，若文献中无此数据，可用下述经验规则推算。系统的压力接近大气压时，压力每降低 1.33 kPa (10 mmHg)，则沸点下降 0.5 ℃。若系统处于较低的压力状态下，压力降低一半，沸点下降 10 ℃。更精确的压力与沸点的关系可以用图 2.3.1 的压力-温度关系图来估算。

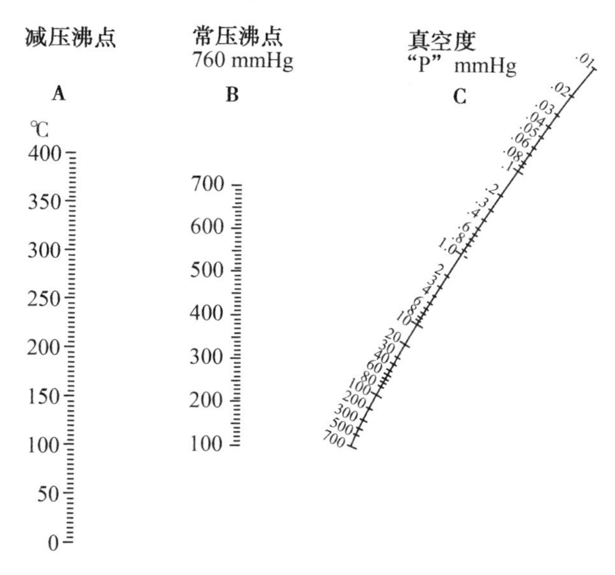

图 2.3.1　液体在常压、减压和真空状态下的压力-温度关系图(1 mmHg≈133 Pa)

(A nomograph used to estimate boiling point of liquid at reduced pressures)

减压蒸馏装置如图 2.3.2 所示，整套装置分为蒸馏、抽气系统、安全保护、接收四部分。减压蒸馏常用克氏(Claisen)蒸馏头，它具有两颈，可避免蒸馏瓶内液体由于暴沸或起泡而冲入冷凝管中。蒸馏烧瓶带支管的磨口插入温度计，另一个磨口插入一根末端拉细的毛细管，毛细管伸至距瓶底 1～2 mm 处。在减压蒸馏时，空气由毛细管进入瓶中，冒出气泡，以防止液体过热而引起暴沸，并使沸腾状态保持平稳，同时又起到一定的搅拌作用。通常在毛细管的上端套一小段橡皮管，并用螺旋夹调节进入烧瓶的空气量，使液体保持适当程度的沸腾。

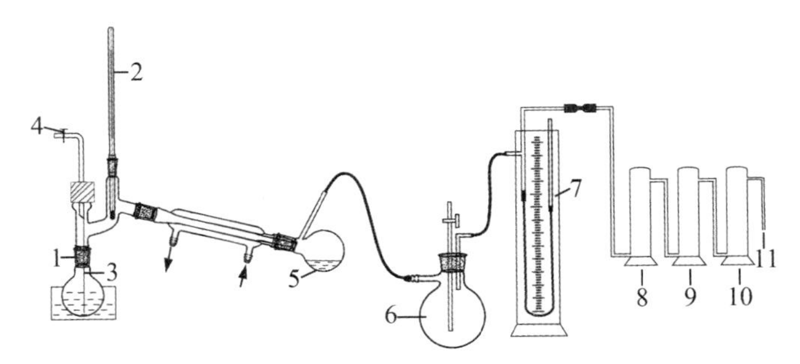

1—蒸馏烧瓶(distilling flask)；2—温度计(thermometer)；3—毛细管(capillary)；4—螺旋夹(screw clamp)；5—圆底烧瓶(round-bottomed flask)；6—安全瓶(flask)；7—水银压力计(manometer)；8—氢氧化钠-碱石灰塔(NaOH-CaO tower drier)；9—无水氯化钙塔($CaCl_2$ tower drier)；10—石蜡屑塔(paraffin tower drier)；11—接真空泵(connect vacuum pump)

图 2.3.2 减压蒸馏装置(Vacuum distillation apparatus)

减压蒸馏装置中的接收器常用蒸馏烧瓶，因为它能耐外压。切不可用平底烧瓶或锥形瓶来作接收器。若要连续收集不同沸点范围的馏出液，则可采用三叉燕尾管。

根据馏出液沸点的不同，选用合适的热浴和冷凝管。不允许在石棉网上直接加热液体，也不能在减压情况下向水浴或油浴中加入冷水或油。热浴的温度较液体沸点高 20～30 ℃。对于蒸馏沸点较高的物质，最好用石棉绳或石棉布包裹蒸馏瓶的两颈，以减少热量的散失。

在实验室内，常用水银压力计来测定减压系统的压力。图 2.3.2 是封闭式压力计，在 U 形管内装入汞，使封闭的一端充满汞。当压力计接到真空装置上时，封闭管内的汞柱开始下降，直到停留于一定高度。此时两管内汞平面之差即减压系统的压力。

抽气泵前面还应连接一个安全瓶(或抽滤瓶)，安全瓶上玻璃管的螺旋夹用来调节压力或放气。抽气泵有油泵和水泵两种。若用油泵，则需在安全瓶和油泵之间装置一系列洗气瓶或吸收塔以吸收水蒸气和其他能腐蚀油泵的气体，保护油泵。若用水泵，则比较方便，但真空度较差。

减压蒸馏系统必须保持密封不漏气，所以常用耐压橡皮管连接。所用橡皮管的大小应与具孔塞上的钻孔配套，具孔塞外层可涂上火棉胶等，干后即可将连接处密封。

【仪器与试剂】

仪器：减压蒸馏装置，油泵或水泵，酸、碱吸收塔，铁架台。

试剂：乙酰乙酸乙酯。

【实验步骤】

在 50 mL 蒸馏烧瓶中加入 20 mL 乙酰乙酸乙酯，按图 2.3.2 装好仪器。旋紧毛细管上的螺旋夹，打开安全瓶上的螺旋夹，然后开泵抽气。逐渐旋紧螺旋夹，通过旋转螺旋夹来调节空气流量，从而达到所需的真空程度。如果因漏气而不能达到所需的真空程度，需检查所有连接部位，塞紧橡皮塞，必要时可涂上火棉胶密封，直到不漏气为止。

仪器装置经检查符合要求后，即可开始蒸馏。开启冷凝管连接的水龙头，然后将水浴加热至沸。蒸馏瓶的圆球部位至少应有 2/3 浸在水浴中，达到每秒钟蒸出 1～2 滴馏出液的效果。在整个蒸馏过程中，应该常注意蒸馏情况，并记录压力、沸点等数据，如有需要可调节螺旋夹，使液体在一定压力下保持平稳沸腾。

蒸馏完毕时，先停止加热，撤去水浴，待稍冷后，慢慢开启安全瓶上的活塞及毛细管上的螺旋夹，使系统内外压力平衡，最后关闭油泵。

【结果记录】

实验完成后，将结果填在表 2.3.1 中。

表 2.3.1　结果记录表

流出量	第一滴	5 mL	10 mL	15 mL
沸点/℃				
真空度/mmHg				
真空度/Pa				

【解释说明】

1. 在减压蒸馏系统中，切勿使用有裂缝的或薄壁的玻璃仪器，尤其不能使用不耐压的平底瓶，以防爆裂。

2. 减压蒸馏时要控制气压稳定，应该将气压调至所需真空度后再加热。

3. 蒸馏结束后，安全瓶的活塞一定要缓慢打开，如果打开太快，系统内外压力突然变化，会使水银压力计的压差迅速改变，导致水银柱破裂。

【操作注意事项】

1. 蒸馏瓶和接收瓶均需使用耐压的圆底厚壁玻璃仪器，防止内外压差过大而引起爆炸。

2. 安装仪器时，应在磨口处涂抹火棉胶，实验开始前，排查实验装置的密封性。

3. 减压蒸馏时，由于液体在较低温度的条件下便可被蒸出，因此不可加热过快。

4. 蒸馏完毕后，应先关闭并撤去热源，慢慢打开放空阀，待压力计读数恢复至零时，再关闭真空泵。

【废物处理】

实验完毕后，蒸馏瓶和接收瓶中的液体均应倒入有机溶剂回收桶中。

【思考题】

1. 什么叫减压蒸馏？怎样的化合物可利用减压蒸馏加以提纯？

2. 要使减压系统达到最大程度的真空，应注意哪些问题？

2.4　水蒸气蒸馏

【实验目的】

1. 学习水蒸气蒸馏的原理及应用。

2. 掌握水蒸气蒸馏装置的安装及操作方法。

【实验原理】

水蒸气蒸馏是用来分离和提纯液态或固态有机化合物的一种方法。当向不溶或难溶于水的有机物中通入水蒸气时，系统内的蒸气压等于水和有机物蒸气压之和，当其值与大气压相等时，混合物沸腾，两者同时被蒸出，这时混合物的沸点低于任何一个组分的沸点。由此可安全地蒸出那些接近或到达沸点时易分解的有机物。

1. 被提纯物质必须具备以下几个条件：

(1)不溶或难溶于水。

(2)共沸时与水不发生化学反应。

(3)在 100 ℃左右时，必须具有一定的蒸气压 0.067～0.133 kPa(5～10 mmHg)，并能随水蒸气挥发。

2. 水蒸气蒸馏常用于下列几种情况：

(1)混合物中含有大量的树脂状杂质或不挥发杂质，采用蒸馏、过滤、萃取等方法难以分离。

(2)某些高沸点的化合物，常压蒸馏会分解、变质、变色等。

(3)从较多的固体反应物中分离被吸附的液体。

常用的水蒸气蒸馏装置如图 2.4.1 所示。水蒸气发生器一般是铁质或铜质的，也可以用圆底烧瓶替代，通常盛水量为其容积的 3/4。安全玻璃管要插到发生器的底部，发挥调节内压的作用。如果水从玻璃管上口喷出，此时应检查整个系统是否堵塞(通常是圆底烧瓶内蒸气导管下口被堵塞)。

水蒸气发生器与圆底烧瓶之间应装一个 T 形管。在 T 形管下端连有一个螺旋夹，以便及时除去冷凝下来的水滴。应尽量缩短水蒸气发生器与圆底烧瓶之间的距离，以减少水蒸气的冷凝。进行蒸馏时，加热水蒸气发生器，直至接近沸腾后再将螺旋夹拧紧，使水蒸气均匀地进入圆底烧瓶。蒸馏需要中断或蒸馏完毕后，一定要先打开螺旋夹通大气，然后方可停止加热，否则烧瓶中的液体会被倒吸到发生器中。在蒸馏过程中，若发现安全管中的水位迅速上升，则表示系统中发生了堵塞。此时应立即打开螺旋夹，然后移去热源，待排除堵塞后再继续蒸馏。

【仪器与试剂】

仪器：水蒸气发生器，蒸馏装置，玻璃弯管，螺旋夹，三角瓶。

试剂：萘。

【实验步骤】

取 4.0 g 萘加到 50 mL 圆底烧瓶中，加水于发生器内，连接好水蒸气蒸馏装置。

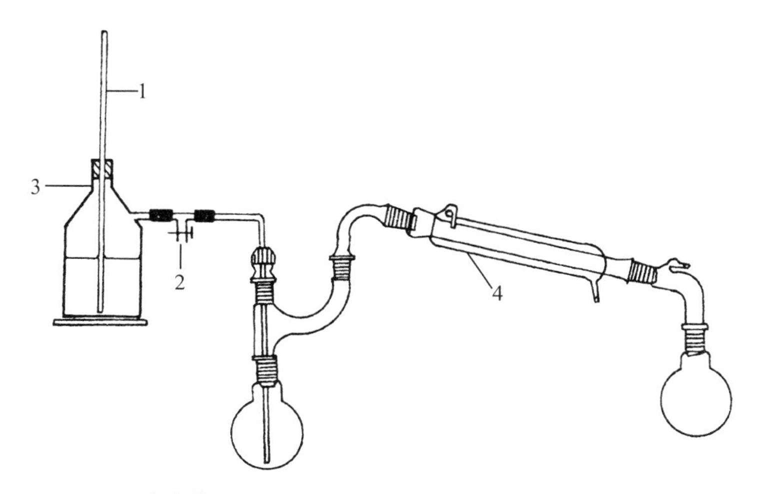

1—安全管(safety glass tube)；2—T形管(type T tube)；

3—水蒸气发生器(water vapor generator)；4—冷凝管(condenser)

图 2.4.1 水蒸气蒸馏装置(Steam distillation apparatus)

加热水蒸气发生器，当水沸腾后，立即关闭 T 形管的螺旋夹，使水蒸气通入烧瓶中。此时，可看到瓶中的混合物翻腾不息，随后在冷凝管中出现有机物和水的混合物。调节火焰使瓶内的混合物不会剧烈飞溅，并控制馏出液的速度为每秒 2～3 滴。待馏出液无油珠，澄清透明时，蒸馏结束。最后，先打开 T 形管的螺旋夹，再移去热源，将产品冷却、抽滤并干燥、测熔点。

【解释说明】

1. 蒸气发生器内水的量不能超过其容积的 3/4，瓶中插一根安全管，蒸馏过程中要适时补加水。

2. 导气管一定要插至接近烧瓶的底部，为防止大量的水蒸气冷凝在烧瓶中，常在烧瓶下方用小火加热。

【操作注意事项】

1. 蒸馏结束，待馏出液冷却后再进行抽滤操作。

2. 若冷凝管中有固体析出,可暂停通冷凝水,使固体熔化后流入接收瓶中。

3. 如果发生倒吸现象,应立即打开螺旋夹,再进行故障排除。

【废物处理】

实验完成后,最终的产物倒入专用回收容器中。

【思考题】

1. 水蒸气蒸馏的原理是什么? 其适用范围是什么?

2. 进行水蒸气蒸馏时,水蒸气导入管末端为什么要插至接近容器底部?

3. 若实验完毕后,先撤掉水蒸气发生器的热源会有何结果?

2.5 折光率的测定

【实验目的】

折光率的测定(操作视频)

1. 了解测定折光率的原理、意义及阿贝折光仪的基本结构。

2. 掌握折光仪的使用方法。

【实验原理】

折光率是物质的特性常数,固体、液体和气体都有折光率,尤其是液体有机化合物。折光率不仅作为物质纯度的标志,也可用来鉴定未知物。如分馏时,配合沸点,作为划分馏分的依据。物质的折光率随入射光线波长不同而变,也随测定时温度不同而改变,通常温度升高 1 ℃,液态化合物折光率降低 $3.5\times10^{-4}\sim5.5\times10^{-4}$,所以折光率($n$)的表示需要注明所用光线波长和测定的温度,常用 n_D^t 来表示,其中 t 表示温度,D 表示钠光的 D 线。例如,水的折光率 $n_D^{20}=1.3330$,即用钠光源(钠光谱中 D 线的波长为 589.3 nm)在20 ℃时测得的水的折光率。

当光线由一种透明介质进入另一种透明介质时,即产生折射现象,见图 2.5.1。此时,入射角(α)的正弦与折射角(β)的正弦之比为常数。光线由介质 A 射入介质 B 时的折射率用数学式表示如下:

$$n=\frac{\sin\alpha}{\sin\beta}$$

通常以空气(介质 A)作为标准介质,当光线由空气进入另一种物质(介质 B)时,折光率为光在空气中的速率和在待测介质中的速率之比,即

$$n=\frac{V_{空}}{V_{液}}=\frac{\sin\alpha}{\sin\beta}$$

由于光线在空气中的传播速度比在液体中的传播速度快,所以液体的折射率总是大于 1。如果介质 A 对于介质 B 是光疏介质,则折射角 β 必小于入射角 α。当入射角 $\alpha=90°$时,$\sin\alpha=1$,这时折射角达到最大值,称为临界角,用 β_0 表示,则

$$n=\frac{1}{\sin\beta}$$

若介质不同,则临界角也不同。根据临界角的大小,由上式便可得出物质的折光率。为了测定临界角,阿贝折光仪采用了"半明半暗"的方法,即让单色光由 0°~90°的所有角度从介质 A 射入介质 B,这时介质 B 中临界角以内的整个区域均有光线通过,因而是明亮的;而临界角以外的全部区域没有光线通过,因而是黑暗的。明暗两区的界线十分清楚。如果在介质 B 上方用目镜观察就可以看见一个界限十分清晰的半明半暗的图像,如图 2.5.2 所示。此时,在显示窗中就可以直接读出该物质的折光率(仪器本身已将临界角换算成折光率)。

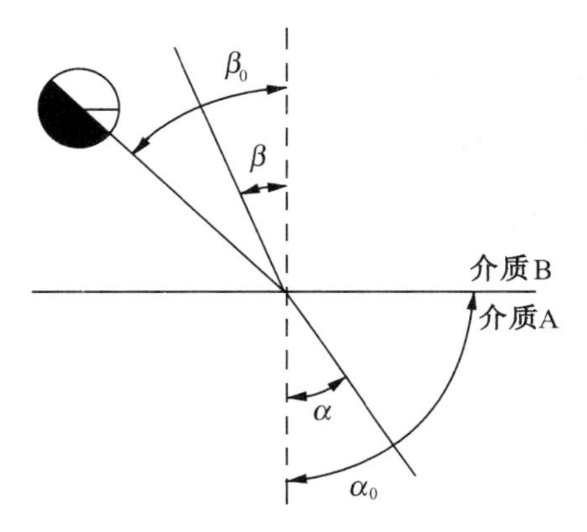

图 2.5.1　光的折射现象(Refraction of light)

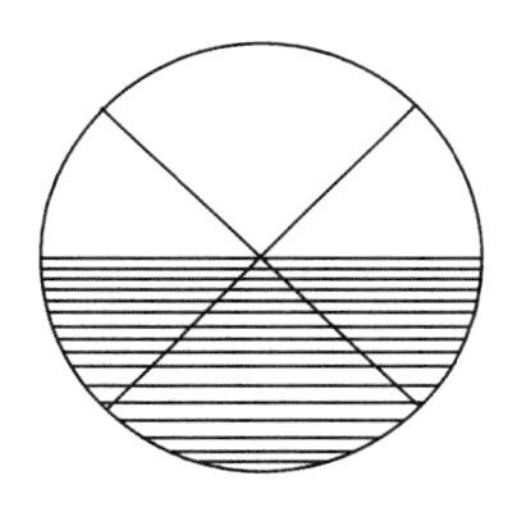

图 2.5.2　阿贝折光仪在临界角时目镜视野图
(Eyepiece view of abbe refractometer at critical angle)

【仪器与试剂】

仪器:WAY-2S(3S)数字型阿贝折光仪。

试剂:95 %乙醇,蒸馏水,待测样品 A、B、C、D。

【实验步骤】

WAY-2S 数字型阿贝折光仪结构图如图 2.5.3 所示。

1. 按下电源开关"POWER"。

2. 打开折射棱镜,检查上、下棱镜表面,并用 95%酒精和擦镜纸小心清洁其表面。

3. 将待测样品放在下面折射棱镜的工作面上。若样品为液体,可用干净滴管吸 2~3 滴液体样品放在棱镜工作面上,然后将上面的进光棱镜轻轻盖上。

4. 转聚光照明部件的转臂和聚光镜筒,使进光表面得到均匀照明。

5. 通过目镜观察视场,同时旋转调节手轮,使明暗分界线落在交叉线交点上(见图 2.5.2)。

6. 旋转目镜下方的色散校正手轮,同时调节聚光镜位置,使视场中明暗两部分具有良好的反差和明暗分界线,尽可能消除色散。

7. 按读数显示键"READ"，显示被测样品的折光率。按温度显示键"TEMP"，显示样品温度。

8. 样品测量结束后，必须用95%酒精和擦镜纸小心清洁棱镜表面，并在两棱镜之间垫上擦镜纸。

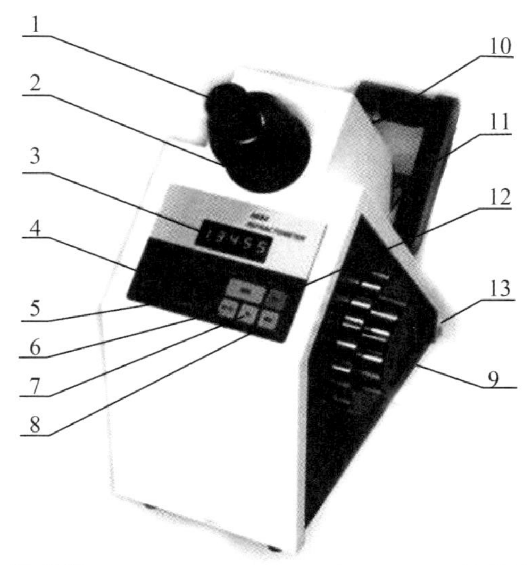

1—目镜(eyepiece)；2—色散手轮(dispersion correction knob)；3—显示窗(display window)；4—"POWER"电源开关(power switch)；5—"READ"读数显示键[reading display button (READ)]；6—"BX-TC"经温度修正锤度显示键[Brix (through temperature correction) display button (BX-TC)]；7—"n_D"折射率显示键[refractive index display button (n_D)]；8—"BX"未经温度修正锤度显示键[Brix (not through temperature correction)]；9—调节手轮(adjustment knob)；10—聚光照明部件(adjustable light)；11—棱镜部件(prisms unit)；12—"TEMP"温度显示键[temperature display button (TEMP)]；13—RS232 接口(RS232 interface)

图 2.5.3 WAY-2S 数字型阿贝折光仪结构图(WYA-2S digital Abbé refractometer)

【结果记录】

实验完成后，将结果填在表 2.5.1 中。

表 2.5.1 结果记录表

待测样品	测量温度	n_D^t				样品名称
		实测值	理论值	校正值	差值	
A						
B						
C						
D						

【解释说明】

1. 可以用蒸馏水对仪器进行校正，折光率的校正公式如下：

$$n_D^{20}=n_D^t+0.00045\times(t-T)$$

式中，t 为实测温度，$T=20$ ℃。

蒸馏水在不同温度下的折光率如表 2.5.2 所示。

表 2.5.2　不同温度下蒸馏水的折射率

温度/℃	折射率 n_D^t	温度/℃	折射率 n_D^t
18	1.33316	25	1.33250
19	1.33308	26	1.33239
20	1.33299	27	1.33228
21	1.33289	28	1.33217
22	1.33280	29	1.33205
23	1.33270	30	1.33193
24	1.33260		

2. 仪器使用前后及更换样品时，必须先清洗擦净折射棱镜系统的工作表面。

3. 如样品为固体，则必须有一个抛光加工过的平整表面。测量前需在折射棱镜工作表面上滴 1～2 滴折射率比固体样品折射率高的透明液体（如溴代萘），不需要将上面的进光棱镜盖上。

【操作注意事项】

1. 被测样品不能有固体杂质。测试固体样品时，应避免折射棱镜的工作表面拉毛或产生压痕，严禁测试腐蚀性较强的样品。

2. 检测有毒物质时，需在通风橱中进行。

3. 使用折光仪时要注意保护棱镜，清洗时只能用擦镜纸而不能用滤纸。

4. 加试样时不能将滴管口触及棱镜表面。

【废物处理】

实验完成后，将实验垃圾倒入专门的回收容器中进行回收处理。

【思考题】

1. 测定折光率有什么意义？其折光原理是什么？

2. 使用折光仪应注意哪些问题？

2.6 旋光度的测定

旋光度的测定

2.7 重结晶和过滤

重结晶和过滤

2.8 液-液萃取

【实验目的】

1. 了解液-液萃取的基本原理。

2. 掌握分液漏斗的基本操作方法。

【实验原理】

萃取是分离和提纯有机化合物的重要操作方法,它的基本原理是利用溶质在两种互不相溶的溶剂中溶解度或分配系数的不同,使溶质从一种溶剂转移到另一种溶剂,经过反复多次的提取,能使大部分溶质提取出来,从而达到分离和提纯的目的。液-液萃取一方面可以从液体混合物中提取所需要的物质,另一方面可用来除去混合物中的少量杂质。分配定律是萃取法的主要理论依据。在一定温度下,当某溶质溶于相互接触而互不相溶的两溶剂时,该溶质在两种溶剂中的浓度比为常数,这个常数被称为分配系数。分配系数 K 可用下式表示:

$$K=\frac{c_{\mathrm{A}}}{c_{\mathrm{B}}}$$

式中,c_{A} 为溶质在A溶剂中的浓度,c_{B} 为溶质在B溶剂中的浓度。

应用分配定律,可计算反复萃取后,剩余溶质的量。设A溶剂有V mL,含溶质W_0,用B溶剂萃取。若用L mL B溶剂第一次萃取后,A溶剂中剩余溶质量为W_1,那么

$$K=\frac{c_A}{c_B}=\frac{\frac{W_1}{V}}{\frac{W_0-W_1}{L}}=\frac{W_1L}{V(W_0-W_1)} \text{或} W_1=W_0\frac{KV}{KV+L}$$

若用L mL B溶剂萃取第二次,A溶剂中剩余溶质量为W_2,那么

$$K=\frac{\frac{W_2}{V}}{\frac{W_1-W_2}{L}}=\frac{W_2L}{V(W_1-W_2)} \text{或} W_2=W_1\left(\frac{KV}{KV+L}\right)=W_0\left(\frac{KV}{KV+L}\right)^2$$

显然经过n次萃取后,A溶剂中剩余溶质的量为

$$W_n=W_0\left(\frac{KV}{KV+L}\right)^n$$

由上式可以看出,n越大,W_n越小。这就说明,把一定量溶剂分成n份,多次萃取比用全部溶剂的量做一次萃取的效果好。

【仪器与试剂】

仪器:60 mL分液漏斗,锥形瓶,100 mL烧杯,碱式滴定管,铁架台(带铁环),25 mL量筒。

试剂:15%醋酸溶液,乙醚,标准氢氧化钠溶液(0.2 mol/L),酚酞指示剂,凡士林。

【实验步骤】

量取5 mL 15%的醋酸溶液倒入60 mL分液漏斗中,再加入14 mL乙醚,塞上玻璃塞,振摇数次,并随时放出分液漏斗内的气体[见图2.8.1(a)],以平衡内部因振摇使乙醚气化所产生的压力。然后静置漏斗于铁环上,取下漏斗的玻璃塞或使玻璃塞的凹槽对准漏斗的小孔,以使萃取体系与大气相通,静置分层[见图2.8.1(b)]。当两层液体完全分层后,慢慢开启下端活塞,放出下层溶液于锥形瓶中,然后向锥形瓶中加5 mL水,并用标准氢氧化钠溶液滴定,用酚酞作指示剂。最后,计算留在水中的醋酸含量,并将上层乙醚液从漏斗上口倒入指定的回收瓶中。

另取5 mL 15%醋酸溶液,先用7 mL乙醚按上述操作方法萃取第一次,然后将下层水溶液分出,置于小烧杯中,上层乙醚液从上口倒入回收瓶中。将烧杯中的醋酸溶液倒入分液漏斗中,再用7 mL乙醚萃取第二次,放出的下层溶液后再加5 mL水,然后用标准氢氧化钠溶液滴定。最后,计算醋酸留在水中的含量,比较一次萃取和两次萃取的结果。

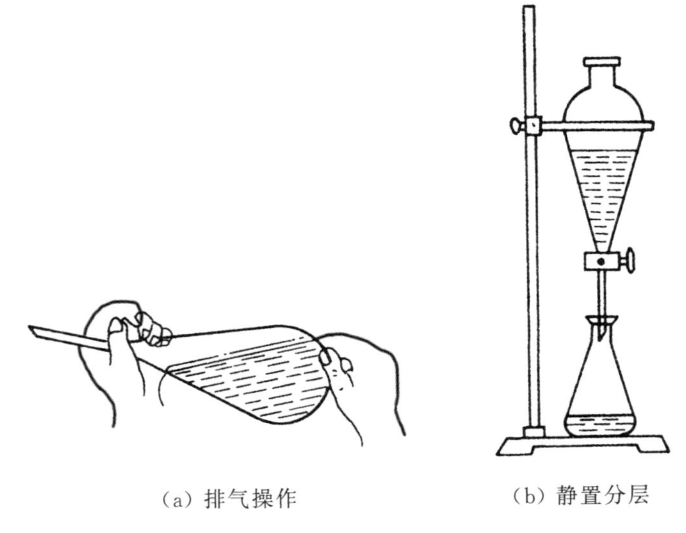

(a) 排气操作　　(b) 静置分层

图 2.8.1　分液漏斗的使用(the use of separatry funnel)

【解释说明】

1. 分液漏斗应选择比萃取液体积大 1～2 倍的，否则会影响萃取物质在两种溶剂中的分配，从而降低萃取率。使用前，应加水检查活塞和玻璃塞是否漏水。

2. 体系静置分层时，必须与大气相通，如图 2.8.1(b)所示，否则会影响分层效果，导致分液时下层液体无法流出。

萃取时，由于剧烈的振摇(尤其有碱性物质或表面活性较强的物质存在时)，常常会产生乳化现象，不能分层或不能很快分层。这种情况下，可采取如下措施：①长时间静置；②在水溶液中加入一定量的电解质(如氯化钠)，利用盐析作用破乳。

【操作注意事项】

1. 漏斗的振摇方法如图 2.8.1(a)所示，用右手手掌顶住漏斗上端玻璃塞，手指握住漏斗的颈部。左手握住漏斗的活塞部分，大拇指和食指按住活塞柄，中指垫在塞座下边。振摇时，将漏斗稍倾斜，下端向上，便于自活塞放气。振摇时，会有大量蒸汽产生，要随时打开活塞放气。放气时，漏斗不能对着自己或别人。

2. 振摇完毕静置分层后，进行分液操作时应先打开漏斗上端玻璃塞，使漏斗内部与大气相通，否则液体不易流出，会造成气泡进入液体内部，导致溶剂重新混合，界面不清楚。

3. 液体分层后，下层液体由下端经活塞流出，上层液体则通过上端磨口倾倒出，以避免造成污染。

4. 液体分层后，应通过液体密度正确判断上下两种液体的成分，如果对液体密度掌握不准确，则分别保留两层液体至实验结束，随时根据实验现象和结果区分所需液体，最后再处理不需要的液体。

【废物处理】

有机层和无机层分别倒入相应回收桶，必要时可借助漏斗倾倒，避免撒漏。若无机溶剂为酸性或碱性溶液，可采用稀溶液进行酸碱中和反应至中性，然后再倒入相应废液桶中。

【思考题】

1. 萃取法的原理是什么？
2. 如何提高萃取效果？如何选择合适的溶剂？
3. 如何采用分液漏斗对互不相容的两种液体进行分离？

2.9 色谱法

色谱法是一种建立在相分配原理上的分离、提纯和鉴定有机化合物的实验方法。利用混合物中各组分在某一物质中的吸附或者溶解性能(即分配)的不同，让混合物的溶液流经该物质，经过反复的吸附或分配等作用，从而将各组分分开，其中流动的体系称为流动相。流动相可以是气体，也可以是液体。固定不动的物质称为固定相，固定相可以是固体吸附剂，也可以是液体(吸附在支持剂上)。根据组分在固定相中的作用原理不同，可以分为吸附色谱、分配色谱、离子交换色谱、排阻色谱等。按操作条件可分为薄层色谱、柱色谱、纸色谱、气相色谱和高压液相色谱等。流动相的极性小于固定相的极性时为正相色谱，而流动相的极性大于固定相的极性时为反相色谱。

2.9.1 柱色谱法

【实验目的】

1. 进一步理解层析法分离、提纯有机化合物的基本原理。
2. 掌握柱色谱法的基本操作技能。

柱色谱法

(实验原理)

柱色谱法

(操作视频)

【实验原理】

柱色谱法(柱层析法)是色谱法的一种，依据其作用原理又可分为吸附柱色谱法、分配柱色谱法和离子交换柱色谱法等，其中吸附柱色谱法应用最广。吸附柱色谱法利用混合物中各组分被吸附能力的不同来进行分离。

柱色谱法通常采用表面积很大并经过活化的多孔性物质或粉状固体作为吸附剂，将其填装入一根玻璃管中，即为固定相，加入待分离混合样品，然后从柱顶加入洗脱剂洗脱。由于化合物中各组分被吸附能力不同，即发生不同解吸，从而以不同速度下移，形成若干色带，若继续再用溶剂洗脱，则吸附能力最弱的组分首先被洗脱出来，如图 2.9.1 所示。整个层析过程进行着反复的吸附—解吸—再吸附—再解吸，使混合物达到分离。分别收集各组分，再逐个鉴定。

常用的吸附剂有氧化铝、硅胶、氧化镁、碳酸钙和活性炭。吸附剂一般要经过纯化和活性处理，颗粒大小要均匀。柱色谱的分离效果与吸附剂的颗粒度有关，通常使用的吸附剂颗粒大小以100～150目为宜。颗粒太粗，溶液流出太快，分离效果不好；颗粒太细，表面积大，吸附能力强，溶液流速慢。供柱色谱使用的氧化铝有酸性、中性和碱性三种。吸附剂的活性取决于含水量的多少，最活泼的吸附剂含最少量的水。氧化铝的活性分为Ⅰ～Ⅴ级。Ⅰ级的吸附作用太强，分离速度太慢，Ⅴ级的吸附作用最弱。

化合物的吸附性与它们的极性成正比，化合物分子中含极性较大的基团时，吸附性也较强，氧化铝对各种化合物的吸附性按以下顺序递减：

酸和碱＞醇、胺、硫醇＞酯、醛、酮＞芳香族化合物＞卤代物、醚＞烯＞饱和烃

溶解样品的溶剂选择也很重要。通常根据被分离化合物中各种成分的极性、溶解度和吸附剂的活性等因素来考虑。

洗脱剂是将被分离物从吸附剂上洗脱下来所用的溶剂，其极性大小和被分离物各组分的溶解性能对分离效果有极大影响。洗脱剂可以是单一溶剂，也可以是混合溶液。一般极性较大的溶剂容易将样品洗脱下来，但为了达到分离的目的，一般使用一系列极性依次增大的溶剂作为洗脱剂。通常先用薄层色谱选择好适宜的溶剂，以节约时间和药品。

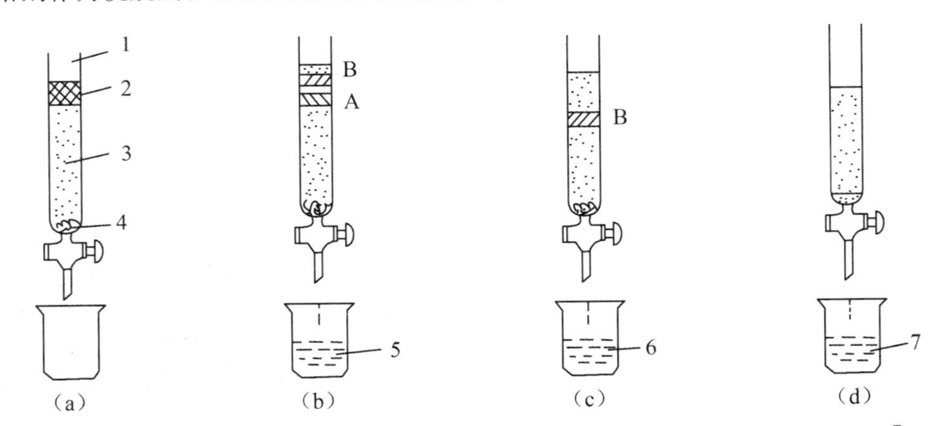

1—色谱柱(chromatographic column)；2—混合样品(A+B)[the mixed samples (A+B)]；3—固定相(stationary phase)；4—棉花(cotton)；5—洗脱液(eluant)；6—洗脱液＋A(eluant＋A)；7—洗脱液＋B(eluant＋B)

图2.9.1 柱色谱的分离过程(The separation process of column chromatography)

本实验采用中性氧化铝作吸附剂来分离荧光素（黄）和亚甲基蓝混合物。氧化铝是一种极性吸附剂，对极性较强的物质（如荧光素）吸附力强，对极性较弱的物质（如亚甲基蓝）吸附力较弱。所以在洗脱过程中，用95％乙醇洗脱时，亚甲基蓝首先被洗脱，荧光素留在色谱柱的上部。洗脱荧光素时，需用极性较大的水。

【仪器与试剂】

仪器：层析柱（色谱柱），150 mL 锥形瓶，60 mL 梨形分液漏斗，小玻璃漏斗，玻璃棒，滴管，普通滤纸，脱脂棉，250 mL 烧杯，100 mL 量筒，25 mL 量筒，100 mL 容量瓶，50 mL 容量瓶，天平，铁架台，蒸馏装置，橡皮管。

试剂：层析用中性氧化铝（100 目），1 g/L 荧光素-亚甲基蓝乙醇混合溶液，95%乙醇，水，海砂。

【实验步骤】

1.装柱

装柱有湿法装柱和干法装柱两种，本实验采用干法装柱。将柱竖直固定在铁支架上，关闭活塞，加入 95%乙醇 15 mL。用一支干净的玻璃棒将少量脱脂棉轻轻推入柱底狭窄部位，小心挤出其中的气泡，但不要压得太紧，否则洗脱剂将流出太慢或根本流不出来。脱脂棉上再铺约 0.5 cm 厚的海沙。打开活塞调节流速为每秒 1 滴，在色谱柱上端通入一个磨口漏斗，慢慢加入 10 g 氧化铝（吸附剂），用套有橡皮管的玻璃棒轻轻敲击柱身，使吸附剂在洗脱剂中均匀沉降，使吸附剂填装均匀紧密。将准备好的海沙加入柱中，使其在吸附柱上均匀沉积成 0.5～1 cm 厚的一层。在装柱过程及装完柱后，都需始终保持吸附剂上面有一段液柱，否则将会有空气进入吸附剂，在其中形成气泡而影响分离效果。如果发现柱中已经形成了气泡，应设法排除。若不能排除，则应倒出重装。柱色谱装置如图 2.9.2 所示。

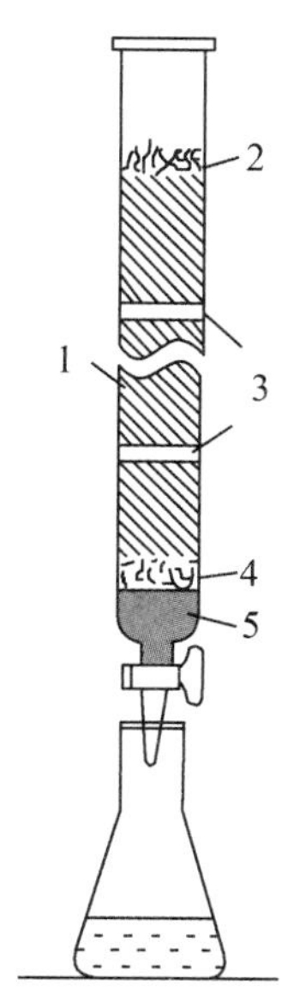

1—吸附剂（adsorbent）；2—海沙（sea sand）；3—样品（sample）；4—海沙（sea sand）；5—脱脂棉（cotton）

图 2.9.2　柱色谱装置图（Schematic diagram of column chromatography）

2.加样和洗脱

当洗脱剂流至离柱顶海沙约 1 cm 时，关闭活塞，加入荧光素-亚甲基蓝乙醇混合溶液 10 滴。开启活塞，当荧光素-亚甲基蓝乙醇混合溶液的液面与柱顶海沙层上面相平时，慢慢加入 95%乙醇，直到观察色带的形成和分离。亚甲基蓝的谱带与被牢固吸附的荧光素谱带分离，继续加足够量的 95%乙醇，使亚甲基蓝从柱子里洗脱下来，洗至洗脱液无色。锥形瓶中的乙醇可通过蒸馏将乙醇蒸去，即可得到纯净的亚甲基蓝。换水作洗脱剂，这时荧光素立刻向柱子下部移动，用锥形瓶收集。

【解释说明】

1. 装柱之前应先将脱脂棉用 95%乙醇润湿，否则柱子里含有气泡。

2. 柱子填装紧密与否，对分离效果影响很大。若柱子留有气泡或各部分松紧不匀（更不能有断层），会影响渗滤速度和显色的均匀。

3. 荧光素和亚甲基蓝能溶于乙醇中，荧光素和亚甲基蓝结构如图 2.9.3 所示。

(a) 荧光素（fluorescein）　　(b) 亚甲基蓝（methylene blue）

图 2.9.3　荧光素和亚甲基蓝的分子结构

（Molecular structure of fluorescein and methylene blue）

【操作注意事项】

为了保持层析柱的均一性，在整个操作过程中应使整个吸附剂润泡在洗脱剂或溶液中。否则，当柱中洗脱剂或溶液流干时，就会发生干裂，影响滤渗和显色的均一性。

【废物处理】

实验完成后，使用过的洗脱剂应倒入专门的回收容器中回收处理。

【思考题】

1. 本实验分离该化合物的依据是什么？

2. 若层析柱中留有气泡或填充不均匀，会怎样影响分离效果？如何避免？

3. 洗脱剂在层析柱中的流速会不会影响柱色谱的结果？有哪些因素需要考虑？

2.9.2　纸色谱法

纸色谱法（实验原理）

纸色谱法（操作视频）

【实验目的】

1. 学习纸色谱法的原理和应用。

2. 掌握纸色谱法分离和鉴定氨基酸的操作技术。

【实验原理】

纸色谱法(亦称纸层析)属于分配色谱法。它以滤纸作载体,以滤纸纤维所吸附的一定量的水分作固定相,以含有一定比例水的一种或多种有机溶剂(亦称展开剂)作流动相,利用滤纸纤维的毛细作用,促使流动相在滤纸上缓缓移动,借助混合物各组分在两相之间分配系数不同,随流动相移动的速度也不同,来使混合物分离。通常,亲水性强的组分随展开剂上升的速度较慢,易于留在固定相中;亲脂性强的组分则上升速度较快。纸色谱装置如图 2.9.4 所示。

通常,我们用比移值(R_f)来表示各组分相对移动速率。比移值是指组分移动距离与溶剂移动距离的比值(见图 2.9.5),其计算公式如下:

$$R_f=\frac{\text{原点至层析点中心的距离}}{\text{原点至溶剂前沿的距离}}=\frac{a}{b}$$

比移值与物质的分子结构、展开剂系统的性质、pH 值、层析温度及层析滤纸的类型等因素有关。在一定的色谱条件下,比移值是物质的一个物理常数,根据比移值可进行混合样品的分离、鉴定。因影响比移值的因素较多,在鉴定时需采用标准样品作对比实验。

纸色谱法主要用于分离、鉴定多官能团或高极性有机化合物,如糖类、氨基酸、生物碱等,其优点是操作简单,所得的色谱图易于保存。纸色谱法的缺点是费时较长,因为展开过程中,溶剂上升速度随着高度的增加而减慢。

【仪器与试剂】

仪器:层析缸(9 cm×18 cm),新华 1 号滤纸条(6.5 cm×14 cm),毛细管,电吹风,喷雾器,镊子,玻璃板(点样板),线绳,铅笔,尺子。

试剂:0.5%脯氨酸水溶液,0.5%亮氨酸水溶液,两种氨基酸的混合液。

展开剂:正丁醇:冰醋酸:无水乙醇:水=4:1:1:2(V/V)混合后的上清液。

显色剂:0.5%茚三酮无水乙醇溶液。

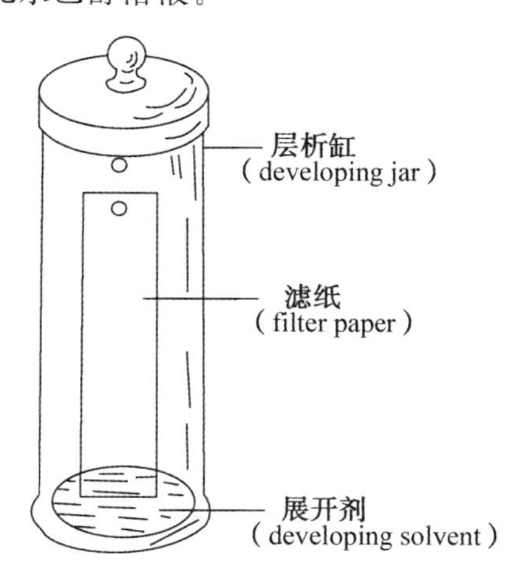

图 2.9.4 纸色谱装置(Paper chromatography)

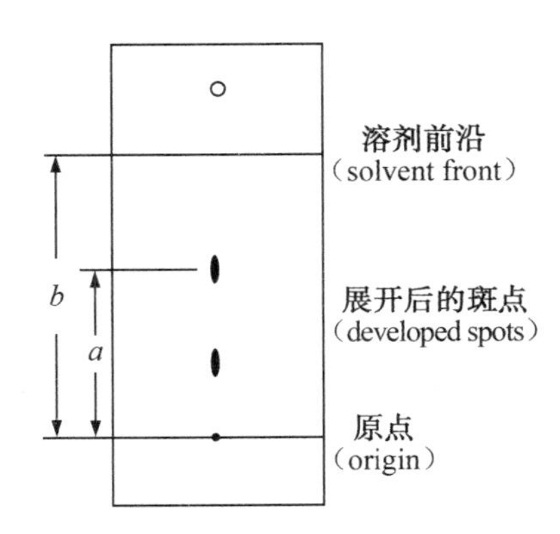

图 2.9.5 比移值的计算(Measurement for the R_f value)

【实验步骤】

1. 点样:取一张滤纸,在滤纸一端 2 cm 处用铅笔画一条横线,在线上等距离处画 3 个点(不可距离滤纸边缘太近),并在滤纸边沿对应于 3 个点用铅笔标上脯、混、亮等字样。取 3 支毛细管,分别蘸取上述氨基酸的水溶液样品,按所标字样分别点上脯氨酸、混合氨基酸及亮氨酸样品。样品点必须尽可能小,每个样品应点样 2~3 次,每次点样后,用电吹风吹干。

2. 展开:层析缸内加入适量展开剂,在滤纸上端打孔处串上绳子,将其挂在层析缸盖内玻璃内勾上,使纸的下缘垂直浸入展开剂中,展开剂液面在原点以下约 1 cm 处(见图 2.9.5),盖上层析缸盖子进行展开,约 1 h 后取出滤纸,并立即用铅笔标出溶剂前沿的位置,用电吹风吹干。

3. 显色:将干燥好的层析纸放在玻璃板上,用喷雾器均匀地将茚三酮乙醇溶液喷于层析纸上,用电吹风吹干,纸条上会显示蓝紫色斑点和黄色斑点,用铅笔画出斑点轮廓。

4.计算比移值与鉴定样品:计算纯样品的比移值以及混合物中两组分的比移值,并与标准的比移值比较,确定混合物中两个斑点的归属。

【解释说明】

1. 层析滤纸应不含杂质,但具有适当的厚度、吸附性和保水性。根据分离样品的种类选择适当的层析滤纸。快速滤纸适用于大多数分离,当需要较好分离效果时,要选择慢速滤纸。

2. 氨基酸溶液尽可能现用现配,并放在冰箱内保存。

3. 展开剂一般根据被分离物质的性质来选择。展开剂应对被分离物质有一定的溶解度,溶解度太大,被分离物质会随展开剂跑到前沿;溶解度太小,被分离物质会留在原点附近,导致分离效果不好。理想的展开剂应该是分配比恒定,不受温度影响,不与分离物质发生化学反应的展示剂。选择合适的展开剂需要查阅相关文献或通过大量实验进行摸索。

4. 样品展开时间取决于被分离物质的性质、展开剂的性质、滤纸的质量和实验温度等因素，比移值控制在 0.4～0.8，分离效果较好。

【操作注意事项】

1. 层析滤纸应用镊子夹取，不能用手拿，因为手指印中含有氨基酸，本实验也能检出。

2. 点样时毛细管要垂直，轻轻触及滤纸即可，点样量要适当。样品量太多，分离效果不好。

3. 展开过程中，样品须高于展开剂液面。

4. 请使用铅笔标记，勿用钢笔或圆珠笔标记。

5. 实验中，随时关注溶剂前沿，应在展开剂到达滤纸上沿之前取出滤纸。

6. 严密盖紧层析缸盖，保持展开剂饱和、蒸气恒定。

【废物处理】

实验完成后，使用过的展开剂应倒入专门的回收容器中进行回收处理。

【思考题】

1. 纸色谱法的原理是什么？比移值的定义是什么？哪些因素会影响比移值？

2. 含有极性不同的多个组分的样品，在极性较小的展开剂中，经纸层析后，能否预测各个组分展开的比移值顺序？解释原因？

2.9.3 薄层色谱法

【实验目的】

1. 了解薄层色谱法的基本原理和应用。

2. 掌握薄层色谱法的基本操作。

【实验原理】

薄层色谱法(薄层层析，简称 TLC)是一种快速分离和定性分析少量物质的重要实验技术，也用于监测反应进程。

薄层色谱法将作为固定相的吸附剂均匀地铺在玻璃板上，制成薄层，然后加上样品，再选择适当溶剂作为流动相(展开剂)，进行展开。由于吸附剂对混合物各组分的吸附能力不同，展开剂带着各组分移动的速度也不同，经过一段时间展开后，各组分最终分离。

通常用比移值表示各组分相对移动速度。比移值是化合物的特征常数，某些化合物在一定条件下的比移值是常数，其数值在 0～1 之间。根据比移值可进行混合样品的分离和鉴定。

常用的薄层色谱法有吸附色谱法和分配色谱法两类。吸附色谱法的吸附剂一般是硅胶和氧化铝，分配色谱法的支持剂为硅藻土和纤维素。

硅胶是无定形多孔性物质，略具酸性，适用于酸性物质的分离和分析。薄层色谱法用的硅胶有以下几类：

硅胶 H：不含黏合剂和其他添加剂。

硅胶 G：含煅石膏黏合剂。

硅胶 HF_{254}：含荧光剂，可在 254 nm 紫外光下观察荧光。

硅胶 GF_{254}：既含煅石膏又含荧光剂等。

氧化铝和硅胶相似，也因含黏合剂或荧光剂而分为氧化铝 G、氧化铝 GF_{254} 和氧化铝 HF_{254}。

氧化铝的极性比硅胶大，比较适于分离极性较小的化合物（烃、醚、醛、酮等）；相反，硅胶适用于分离极性较大的化合物（羧酸、醇、胺等）。

氨基酸的薄层色谱利用硅胶薄层色谱板中的硅胶能与极性基团之间形成的氢键而产生吸附作用。各种氨基酸极性程度不同，被吸附能力的大小也不同。因此，在层析时，各种氨基酸的移动速度也不同。经过一定时间的展开后，氨基酸彼此分离。取出薄层板，待展开剂挥发后，以茚三酮显色，测定比移值。

【仪器与试剂】

仪器：层析缸，玻璃板（10 cm×3 cm），毛细管，电吹风，喷雾器，镊子，线绳，铅笔，尺子。

试剂：硅胶 G，95%乙醇，0.1%精氨酸，0.1%丙氨酸，待分离溶液（丙氨酸＋精氨酸），0.1%羧甲基纤维素钠，展开剂（正丁醇：醋酸：水＝12：3：5），显色剂（0.5%茚三酮丙酮溶液）。

【实验步骤】

1. 制板

称取 2.5 g 硅胶 G 于小烧杯中，加入 0.1%羧甲基纤维素钠水溶液 8 mL，调匀，倒在干净的玻璃板上；然后，倾斜转动玻璃板，使支持物在玻璃板上形成均匀的薄层；将玻璃板平放在桌面上，放置 10～15 min 后放在 110～120 ℃烘箱中 30 min，使其活化；冷却后置于干燥器中备用。

2. 点样

在距薄层板底部 2 cm 处用铅笔轻轻画一条横线，在线上等距离地轻轻点上 3 个点，然后用毛细管分别吸取丙氨酸、精氨酸以及待分离溶液点在这 3 个点上，每个点点 2～3 次（每点一次要等干燥后再点下一次）。

3. 展开

在一干净的层析缸中加入适量展开剂，将展开剂在层析缸中饱和约 10 min，将点有样品的层析板的一端斜放于层析缸中（注意切勿使样品浸入展开剂中），立即盖严，使样品在密闭的层析缸中展开，薄层色谱装置如图 2.9.6 所示。待展开剂前沿到达薄层板上端 2/3 处时，取出薄层板，立即划出溶剂前沿的位置，将层析板自然晾干或用电吹风吹干。

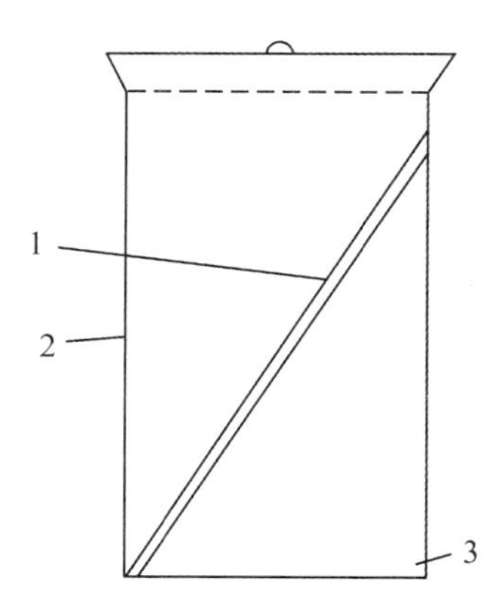

1—层析板(chromatography plate);2—层析缸(chromatography cylinder);3—展开剂(developing agent)

图 2.9.6 薄层色谱装置(Schematic diagram of thin layer chromatography)

4. 显色

在层析板上喷茚三酮溶液,用电吹风吹干,即显出紫色斑点。测量原点到斑点中心的距离及原点到溶剂前沿的距离,分别计算样品的比移值。

【解释说明】

1. 用于制备薄层色谱的玻璃板要求面平、干净,否则硅胶层容易脱落。

2. 薄层板的活性与含水量有关,其活性随含水量的增加而降低。因此,制备好的薄层板在室温下晾干后,还要放在烘箱内加热活化,进一步去除水分。活化条件要根据具体情况而定。一般来说,硅胶板在 105~110 ℃烘干 30 min,氧化铝板在 200~220 ℃烘干 4 h。当分离一些易吸附的化合物时,也可以不活化。

3. 加入展开剂的量以不没及薄板样品为宜。

【操作注意事项】

1. 层析薄层板时应保护好硅胶层,用铅笔画线时不能破坏硅胶层。

2. 点样时毛细管要垂直,轻轻触及薄层板即可,点样量要适当。样品量太多,分离效果不好。

3. 展开过程中,样品需高于展开剂液面。

4. 请使用铅笔标记,勿用钢笔或圆珠笔标记。

5. 严密盖紧层析缸盖,保持展开剂饱和、蒸气恒定。

【废物处理】

实验完成后,使用过的展开剂应倒入专门的回收容器中进行回收处理。

【思考题】

1. 薄层色谱法的原理是什么?

2. 层析时,若加入展开剂没过样品斑点会有什么结果?

3. 在一定的层析条件下,为什么可根据比移值来进行被测样品的定性分析?

2.10 氨基酸的纸上电泳

氨基酸的纸上电泳（实验原理）

氨基酸的纸上电泳（操作视频）

【实验目的】

1. 了解纸上电泳的基本原理。
2. 掌握纸上电泳分离、鉴定氨基酸的操作方法。

【实验原理】

电泳技术较色谱技术发展略晚，是一种与色谱技术结合的方法，亦称电泳色谱法。

电泳是指分散在介质中的带电粒子在电场的作用下，向着与其电性相反的电极方向移动的现象。在一定条件下（如电场强度、pH 值），样品中各组分泳动的速度与本身所带电荷的性质、电荷的数量及相对分子质量有关。因而在同一电场中，各组分泳动的方向和速率不同，在一定的时间内各自移动的距离也不同，从而可达到对某些物质的分离和鉴定的目的。带电质点在电场中的移动速率除了与自身性质有关，还受电场强度、溶液 pH 值、离子强度和电渗等因素影响。

根据有无支持物，电泳可分为自由电泳（无支持物）和区带电泳（有支持物）两种。前者有显微电泳、等速电泳等，后者有纸电泳、薄层电泳、凝胶电泳、毛细管电泳等。区带电泳操作简便，分离效果好，因此常用于分离鉴定。

纸上电泳是以滤纸作为支持物，带电质点在滤纸上受一定的电场作用而移动，从而达到分离目的。样品点在滤纸条上，滤纸用缓冲液浸润，使其能导电，并在电泳过程中保持一定 pH 值不变，将滤纸放在一个支架上，滤纸的两端浸在缓冲液中。接通电源，滤纸的两端就有一定的电压吸引荷电物质在纸上移动。纸上电泳可用于氨基酸、蛋白质、糖类、有机酸、无机离子、配位化合物等物质的分离与鉴定。

各种氨基酸都有其特定的等电点。在等电点时，氨基酸分子本身呈电中性，在直流电场中既不向负极移动也不向正极移动；如果将氨基酸置于 pH 值大于其等电点的溶液中，氨基酸带负电，在直流电场中向正极移动；如果将氨基酸置于 pH 值小于其等电点的溶液中，氨基酸带正电，在直流电场中向负极移动。带电氨基酸移动方向如图 2.10.1 所示。

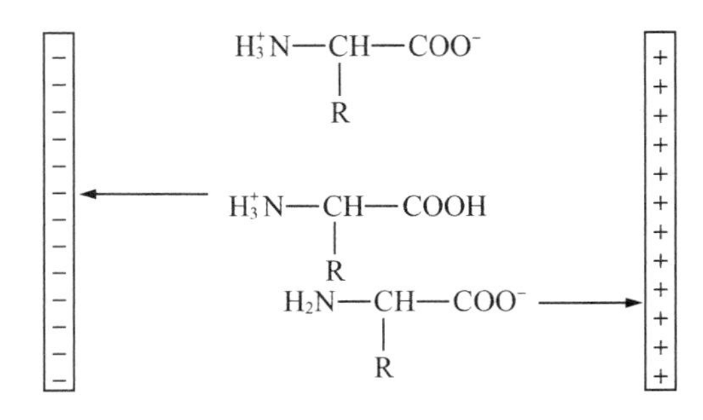

图 2.10.1 带电氨基酸移动方向(Migration direction of charged amino acids)

由于氨基酸混合物中各种氨基酸分子量不同,等电点不同(在一定 pH 值时荷电性质及荷电量不同),因此在同一电场作用下,各种氨基酸泳动的方向和速度必然不同。电泳一段时间后,各种氨基酸在滤纸被分离开。在电泳中,如果溶液的 pH 值距氨基酸的等电点越远,氨基酸移动越快,反之则越慢。对于本实验所用氨基酸,谷氨酸的等电点为3.22,丙氨酸为 6.02,精氨酸为 10.76。在 pH 值为 5.9 的缓冲溶液中,三种氨基酸在电场作用下的移动方向和距离不同,从而实现分离目的。

纸上电泳与纸上色谱一样,常在相同的实验条件下,用标准品作对比实验来鉴定化合物。

【仪器与试剂】

仪器:DYY-4C 电泳仪,滤纸(中华 3 号,4 cm×15 cm),毛细管,喷雾器,镊子,电吹风。

试剂:待分离溶液(丙氨酸、精氨酸、谷氨酸混合液),0.5%茚三酮溶液,邻苯二甲酸氢钾-氢氧化钠缓冲溶液(pH 值为 5.9)。

【实验步骤】

1. 点样和湿润:取滤纸条一张,用铅笔在中央划一横线后,在滤纸两端标好正负极,用毛细管点上混合溶液的斑点,用电吹风吹干后,将滤纸条放在电泳槽的支架上,两端浸入电泳槽的缓冲溶液中(每槽各加入 100~150 mL 缓冲液),当滤纸湿至距样品约 1 cm 处取出,沿水平方向拉直,等待滤纸完全被缓冲液浸润。

2. 电泳:盖好电泳槽,通直流电,调节电泳仪的输出电流强度控制器,电压表先调至 100 V,然后每分钟升 10 V,直至电压为 220 V 后保持不变,泳动 1 h 后停电。

3. 显色:用镊子取出滤纸,用电吹风吹干,用喷雾器喷上茚三酮溶液,再用电吹风吹干或放在 100 ℃烘箱中加热几分钟,滤纸上便显出各种氨基酸的紫色斑点,用铅笔将其圈出。鉴定各组分,并在图 2.10.2 中画出电泳结果。

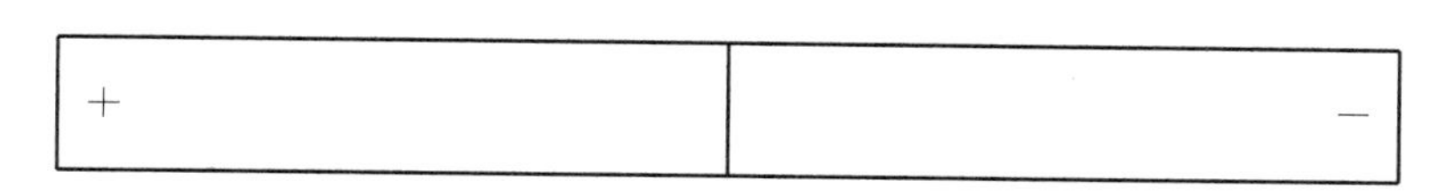

图 2.10.2 电泳结果

【解释说明】

1. 一个带电质点在电场中移动的速率除了与其本身的性质有关外,还与下列因素相关:

(1)电场强度(V/cm):对电泳起着重要作用。电场强度越高,带电质点移动越快。因此,电泳可分为高压电泳(大于 50 V/cm)、常压电泳(10~50 V/cm)和低压电泳(小于 10 V/cm)。

(2)溶液的 pH 值:溶液的 pH 值对某些分子(如氨基酸、蛋白质等两性电解质)的带电状态有较大的影响。

(3)溶液的离子强度：电泳时溶液的离子强度越强，电泳速率越慢；如果离子强度过低，缓冲溶液的缓冲容量小，不宜维持 pH 值的稳定。一般适宜的离子强度为 0.02～0.2 mol/kg。

(4)电渗：在电场作用下，液体对于固体支持物的相对移动称为电渗。由于电渗现象常与电泳同时存在，因而对带电粒子的移动距离有影响。若电泳方向与电渗方向相同，则电泳速率加快；反之，则电泳速率减慢。因此，尽量选择电渗作用小的物质做支持物。

(5)其他因素：缓冲溶液的黏度、缓冲溶液与带电粒子的相互作用以及电泳时的温度变化等因素也都能影响电泳速度。

2. 纸上电泳所分离的物质只限于荷电物质，所以应用范围不如纸层析法广泛。但有些物质，尤其是大分子化合物，用纸色谱法不如电泳法简便，所以纸上电泳法也成为常用的分析方法之一。在医学领域，纸上电泳常用于蛋白质及氨基酸的分离鉴定。

3. pH 值为 5.8 的邻苯二甲酸氢钾-氢氧化钠缓冲液的配制：准确称取 5.10 g 邻苯二甲酸氢钾和 0.86 g 氢氧化钠，加蒸馏水稀释至 1000 mL。

【操作注意事项】

1. 电泳时，因有电流通过滤纸，会产生一定热量，所以应将电泳槽密闭，以防水分蒸发，导致滤纸变干而不能导电。

2. 电压稳定才能得到重现性好的结果，所以电压要逐渐升高。

3. 在整个操作过程中应尽量避免手与滤纸接触，以免显色后指纹太多，影响观察结果。

4. 为防止触电，一定要先切断电源，再取出滤纸，且电泳过程中不可用手或镊子接触滤纸。

【废物处理】

实验完成后，使用过的电泳液和沾有化学试剂的滤纸应分别放入专门的回收容器中进行回收处理。

【思考题】

1. 什么是氨基酸的等电点？等电点与氨基酸分子的带电状态有什么关系？

2. 写出丙氨酸、精氨酸、谷氨酸在 pH 值为 7 的缓冲溶液中的结构式，判断电泳时它们会向哪极移动？

2.11 波谱技术

波谱技术

2.12 模型作业

【实验目的】

模型作业
（实验原理）

1. 观察有机化合物分子的立体结构，掌握有机分子立体结构的书写方式。

2. 掌握同分异构体、构造异构体和立体异构体的概念、分类及命名方法，探讨有机化合物的结构与性质的关系。

【实验原理】

异构现象在有机化学中极为普遍，常见的异构现象如图 2.12.1 所示。

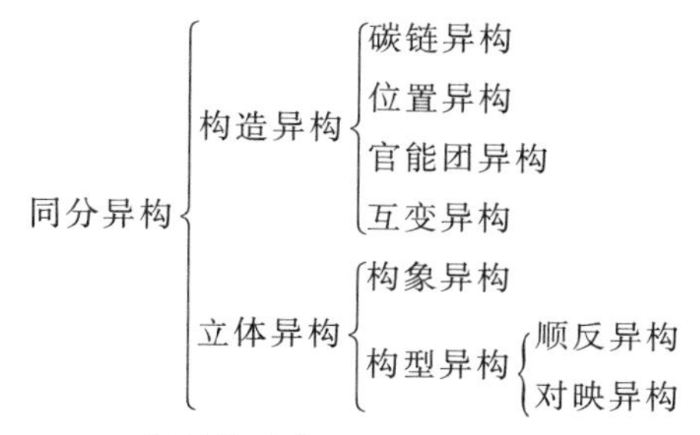

图 2.12.1 常见的异构现象（Common isomerism）

构造异构是因分子中原子或基团的连接顺序或连接方式不同而产生的异构现象。立体异构是指分子中原子或基团的连接顺序或连接方式相同，但由于它们在空间的排列方式不同而产生的异构现象。单键的自由旋转是产生构象异构的主要原因。通常构象异构体的相互转化可以通过碳碳单键的旋转实现，不涉及价键断裂，因此所需能量较低，室温下即可完成。一般条件下，实验不能分离得到纯的构象异构体。构型异构体的相互转化涉及价键的断裂，所需能量高，因此构型异构体在室温下能稳定存在，能够分离获得纯的构型异构体。手性是产生对映异构现象的根源，不能与其镜像重合的分子称为手性分子，手性分子中存在对映异构现象。

分子的构造用构造式表示，如图 2.12.2 所示。构造式常用实线式（Kekulè 式）、简写式和键线式表示。化合物的立体结构有楔形透视式、锯架（透视）式、纽曼投影式和费歇尔投影式四种书写方式。对于楔形透视式，细实线“ — ”表示价键位于纸平面上，楔形虚线“ıııııı”表示价键朝向纸后，楔形实线“▬”表示价键朝向纸前，以此表示化合物的空间立体形象。

$CH_3CH_2CH_2CH_3$

实线式（straight-line formula） 结构简式（condensed formula） 键线式（bond-line） 楔形透视式（wedge perspective）

图 2.12.2 分子的构造式（Constitutional formula）

构象异构一般用锯架（透视）式和纽曼（Newman）投影式表示，如图 2.12.3 所示。锯架式实际上是球棍式模型的简化写法，所有的键都用实线表示，从 C—C 键轴斜 45°方向看，每个碳原子上的其他三根键夹角均为 120°；纽曼投影式是把分子的球棍式模型沿 C—C 键的轴线投影观察得到的。

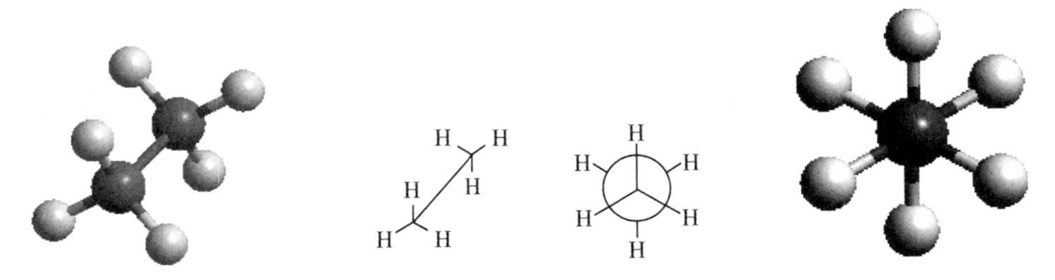

(a)锯架式(sawhorse perspective)　　(b)纽曼投影式(Newman projection)

图 2.12.3　构象异构(Conformational isomers)

在立体化学中，尤其是糖类化学中，费歇尔（Fischer）投影式是表示手性分子立体构型的常用方法，书写 Fischer 投影式的基本原则：①将主链竖置，把编号最小的碳原子（或氧化态较高的碳原子）放在最上端。②按照“横前竖后”原则，与手性碳原子相连的两个横键伸向前方，两个竖键伸向后方。③横线与竖线的交点代表手性碳原子。左旋乳酸的构型可表示为图 2.12.4 所示的形式。

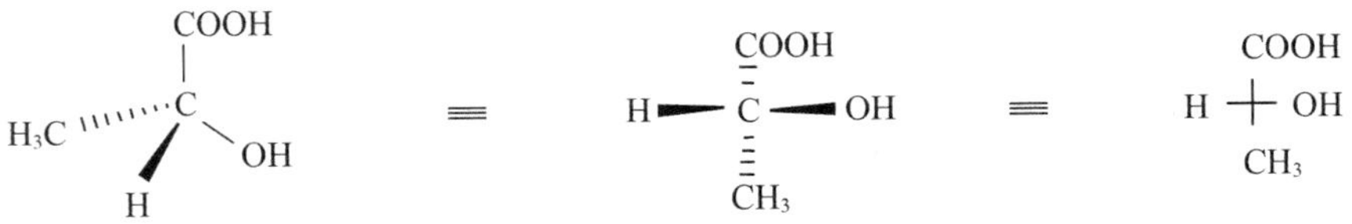

图 2.12.4　(—)-乳酸的 Fischer 投影式(The Fischer Projection of (—)-lactic acid)

糖类化合物的环状结构一般用哈武斯（Haworth）式表示，如图 2.12.5 所示。

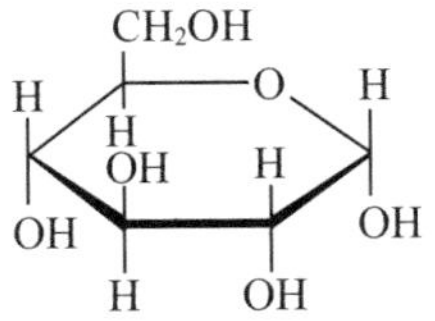

图 2.12.5　α-D-葡萄糖的环状结构(Ring structure of α-D-glucose)

分子结构通常用分子模型来表示，通常使用的一种模型是球棒模型，用不同颜色的小球代表不同的原子或原子团。

例如：

碳 C　黑色　　氢 H　白色　　氧 O　红色　　硫 S　黄色

氮 N　天蓝色　　磷 P　紫色　　氯 Cl　绿色

各种色球间用短棒、套管或弹簧相连接，以表示原子或原子团间的化学键。碳原子相互间通过单键、双键和叁键连接成链或环。

【模型材料】

球棒模型一套。

【模型练习】

1. 搭出甲烷、乙烯和乙炔的分子模型，观察和比较单键、双键、叁键及其键角。画出甲烷、乙烯和乙炔分子的立体结构式(楔形透视式)。

2. 搭出乙烷、丙烷、正丁烷和异丁烷的分子模型。

(1)观察和比较正丁烷、异丁烷的分子模型，了解甲基，乙基，丙基，异丙基以及伯、仲、叔碳原子。

(2)将正丁烷和异丁烷模型变成正丁基、异丁基、仲丁基和叔丁基，写出以上基团的构造式(实线式)。

(3)在乙烷分子模型中将 C—C 单键旋转 360°，观察其构象，从这些构象中了解重叠式、交叉式构象，并用锯架式和 Newman 投影式表示它们。

(4)在正丁烷分子模型中将 C2—C3 单键旋转 360°，观察其构象，分别用锯架式和 Newman 投影式表示出正丁烷的对位交叉式、邻位交叉式、部分重叠式和完全重叠式四种典型的构象异构体，并指出优势构象。

3. 搭出丁烯的所有异构体的模型，并用顺、反或 Z、E 构型命名法命名 2-丁烯。画出两种立体结构式(透视式)，并观察两个甲基的相对远近。

4. 搭出丁烯二酸的模型，确定其顺、反异构体并画出其立体结构式。

5. 搭出乳酸、酒石酸和葡萄糖的分子模型。

(1)观察乳酸分子模型。画出乳酸分子的透视式和 Fischer 投影式，了解 D-乳酸和 L-乳酸及 R-乳酸和 S-乳酸的空间构型及构型命名原则。

(2)观察酒石酸的分子模型。画出酒石酸的三种立体异构体的 Fischer 投影式，并找出内消旋酒石酸分子的对称因素(如对称面或对称中心)。

(3)观察葡萄糖的开链结构和环状结构的分子模型，画出葡萄糖分子开链结构的 Fischer 投影式和环状结构的哈武斯(Haworth)式，了解开链结构与环状结构间的转变关系，并比较 α-D-葡萄糖与 β-D-葡萄糖两种构象的稳定性。

6. 搭出 2-溴-3-氯丁烷的所有立体异构体，画出其 Fischer 投影式并标记构型。

7. 试用分子模型说明下列两组中化合物之间的关系(相同化合物、对映体、非对映体)，并用 R、S 构型命名法标出各个化合物的构型。

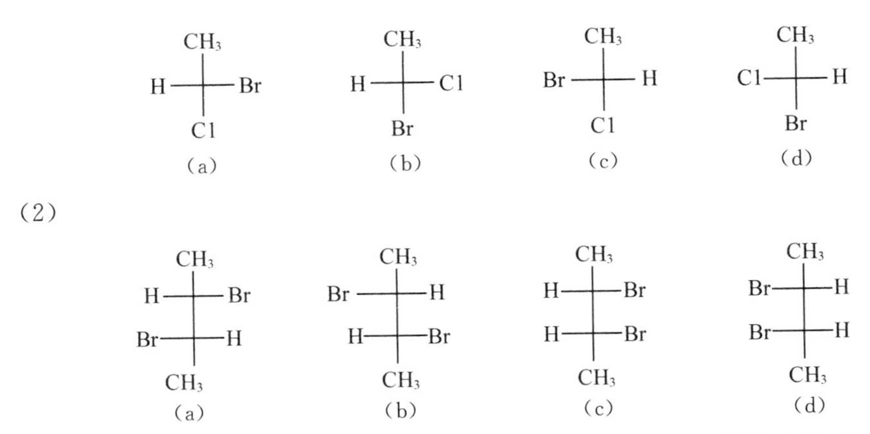

8. 试用模型找出下列化合物(a)～(f)中哪些互为相同构型的化合物，并用 R，S 构型命名法命名。

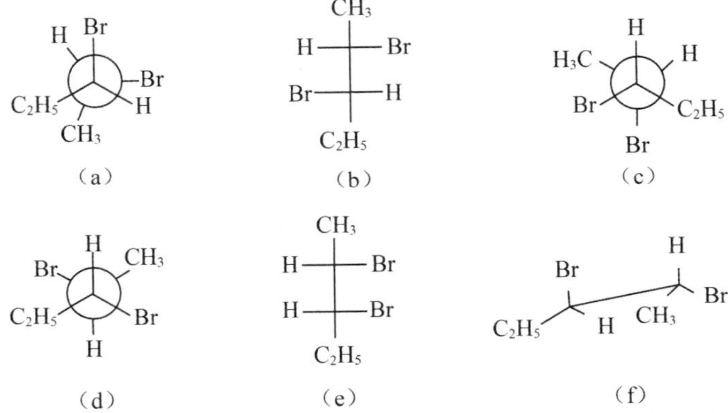

9. 搭出环己烷的分子模型，比较船式构象和椅式构象的稳定性，画出椅式构象的透视式。

10. 下列为甲基环己烷的两种构象式，试以模型说明哪一种较稳定？

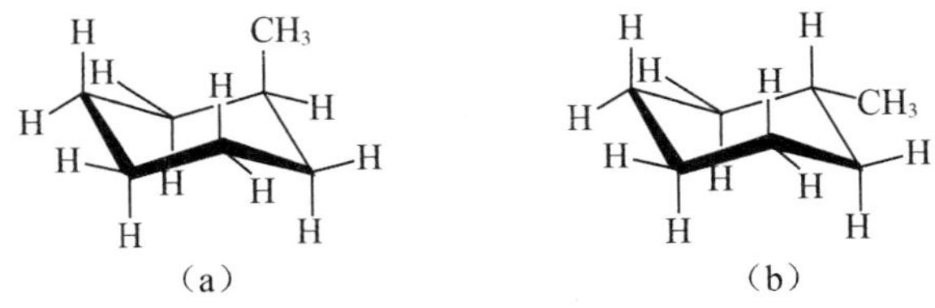

11. 分别搭出 1,4-二甲基环己烷的分子模型，了解其顺、反异构体，比较其稳定性，并画出其椅式构象。

12. 画出反式 1-叔丁基-4-甲基环己烷（2017 年有机化合物命名原则）的稳定构象式，并用模型说明理由。

13. 搭出顺式和反式十氢萘的分子模型。十氢萘可以看成二个环己烷的稠合，环己烷的优势构象是椅式构象。二个椅式构象的稠合方式有两种，一种是 ee 稠合，一种是 ea 稠合。从模型中理解这两种稠合方式是顺式还是反式，并比较两种异构体的稳定性。

【思考题】

1. 请解释构型和构象的概念并指出它们的异同点。
2. 直链烷烃为什么呈锯齿状？烷烃的同分异构体中，支链越多，沸点越低吗？
3. 试说明有机化合物分子中手性碳原子、对映异构现象、手性分子三者之间的关系。

2.13　有机化合物官能团的定性反应

有机化合物官能团的定性反应

第3章　有机化合物的制备

3.1　环己烯的制备

【实验目的】

1. 了解由醇脱水制备烯烃的方法。

2. 掌握分馏及蒸馏的基本操作。

【实验原理】

环己烯通常由环己醇在浓硫酸或浓磷酸催化下脱水制备，由于浓硫酸在反应中易发生炭化，本实验以浓磷酸作脱水剂制备环己烯，其反应式如下：

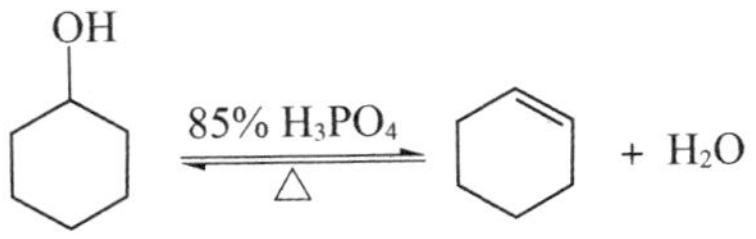

$$\text{环己醇(OH)} \underset{\triangle}{\overset{85\%\ H_3PO_4}{\rightleftharpoons}} \text{环己烯} + H_2O$$

为了促进可逆反应平衡向右移动，本实验采用边反应边蒸出反应生成的环己烯和水形成的二元共沸物(沸点 70.8 ℃，含 10%水)的方法。但是原料环己醇也能和水形成二元共沸物(沸点 97.8 ℃，含水 80%)，故本实验采用分馏装置，并控制柱顶温度不超过90 ℃。

【仪器与试剂】

仪器：圆底烧瓶，分馏柱，加热套，温度计，锥形瓶，烧杯，分液漏斗，蒸馏接头，冷凝管，量筒。

试剂：环己醇，85%磷酸，精盐，5%碳酸钠溶液，无水氯化钙。

【实验步骤】

在 50 mL 的圆底烧瓶中，加入 10.4 mL 环己醇，4 mL 85%的磷酸和 1 粒搅拌磁子。将烧瓶放入磁力搅拌电热套中，开启搅拌使反应物混合均匀。烧瓶上放置短分馏柱，接上冷凝管，用 50 mL 的锥形瓶作接收器，并置于冰水浴中，以免环己烯挥发而损失。

打开电热套加热控制开关，调节电压慢慢加热反应混合物至沸腾，控制分馏柱顶部的温度不要超过 90 ℃，缓慢蒸出生成的环己烯和水。当烧瓶中只剩下少量的残留物时，停止加热。

将馏出液用 1 g 精盐饱和，并轻轻地振摇烧瓶。将馏出液倒入一分液漏斗中，加入 3～4 mL 5%碳酸钠溶液（或用 0.5 mL 20%的氢氧化钠溶液）。振摇后静置分层，分去水层，上层的粗产物从分液漏斗的上口倒入一个干燥的锥形瓶中。加入 1～2 g 无水氯化钙，振荡，直至溶液变澄清。

将干燥后的环己烯滤入一个干燥的 50 mL 圆底烧瓶中，加入 1 粒搅拌磁子，用水浴加热蒸馏。用一称重的 50 mL 圆底烧瓶作接收器，将烧瓶置于冰水浴中。收集 80～85 ℃的馏分，称重产品，计算产率。

纯的环己烯为无色液体，沸点为 82.98 ℃，$n_D^{20}=1.4465$，$d_4^{20}=0.808$（d_4^{20} 指物质在 20 ℃时对 4 ℃水的相对密度）。

【解释说明】

1. 环己烯的沸点为 82.98 ℃。环己烯与水共沸时，沸点为 70.8 ℃。环己醇也能与水形成共沸物，沸点为 97.8 ℃。因此，蒸馏温度需仔细控制，使其不超过 90 ℃。

2. 加精盐的目的是减少产物在水中的溶解度，促进水层与有机层分离。

3. 少量的磷酸也会与产物共沸，故需用碳酸钠水溶液洗涤除去。

4. 在精制产品时，蒸馏所用的仪器需干燥。

【废物处理】

产物倒入专用回收瓶中，水溶液（残渣和洗涤液）用水稀释并中和后冲入下水道，氯化钙可以放置在非危险固体废物容器中。

【思考题】

1. 在制备过程中为什么要控制分馏柱顶部的温度？

2. 在粗制的产品中加入精盐的目的是什么？

3. 为什么反应混合物中存在的少量磷酸必须在蒸馏环己烯前用碳酸钠中和除去？

3.2 1-溴丁烷的制备

【实验目的】

1. 学习用溴化钠、浓硫酸和正丁醇制备 1-溴丁烷的原理和方法。

2. 掌握带气体吸收装置的回流加热及液体化合物的洗涤等基本操作。

【实验原理】

1-溴丁烷是由正丁醇与溴化钠、浓硫酸共热制得的，主反应如下：

$$NaBr + H_2SO_4 \longrightarrow HBr + NaHSO_4$$

$$n\text{-}C_4H_9OH + HBr \xrightleftharpoons{\triangle} n\text{-}C_4H_9Br + H_2O$$

副反应如下：

$$CH_3CH_2CH_2CH_2OH \xrightarrow[\triangle]{H_2SO_4} CH_3CH_2CH{=}CH_2 + CH_3CH{=}CHCH_3$$

$$2CH_3CH_2CH_2CH_2OH \xrightarrow[\triangle]{H_2SO_4} CH_3CH_2CH_2CH_2OCH_2CH_2CH_2CH_3 + H_2O$$

$$2HBr + H_2SO_4 \xrightarrow{\triangle} Br_2 + SO_2\uparrow + 2H_2O$$

【仪器与试剂】

仪器：圆底烧瓶，冷凝管，加热套，温度计，锥形瓶，烧杯，分液漏斗，蒸馏接头，量筒，漏斗。

试剂：正丁醇，溴化钠，浓硫酸，10%氢氧化钠溶液，无水氯化钙。

【实验步骤】

在 100 mL 的圆底烧瓶中，加入 10 mL 水，再小心加入 15 mL 浓硫酸，混合均匀后，冷却至室温。依次加入 7.4 g (9.2 mL，0.10 mol)正丁醇及 12.3 g(0.12 mol)研细的溴化钠粉末。充分振摇后，加入 1 粒磁子，装上回流冷凝管，在冷凝管上端接一气体吸收装置[参考图 1.4.3(b)和图 1.4.6(a)]，用 5%的氢氧化钠水溶液作吸收剂。

用加热套进行加热，使反应液保持微沸，回流约 0.5 h。冷却后，改作蒸馏装置。加热套加热至馏出液完全溶于水(或变为澄清)，此时 1-溴丁烷已全部蒸出，停止蒸馏。

将馏出液小心转入分液漏斗中，用 15 mL 水洗涤，并静置分层。将有机层转入另一个干燥的分液漏斗中。有机层用 6 mL 浓硫酸洗涤，并尽量分去下层硫酸层。有机层依次用 15 mL 水，15 mL 10%氢氧化钠溶液和 15 mL 水洗涤，用 pH 试纸检验是否已达中性，否则重复水洗。

将产物转入一个干燥的锥形瓶中，加入 1～2 g 无水氯化钙干燥，间歇摇动至液体透明。将干燥后的产物转入干燥的 50 mL 蒸馏烧瓶中，并用加热套加热蒸馏，收集 99～103 ℃的馏分。称重产物，计算产率。

纯的 1-溴丁烷为无色透明液体，沸点为 101.6 ℃，$n_D^{20}=1.4401$，$d_4^{20}=1.2758$。

【解释说明】

1. 在回流过程中，尤其是停止回流时，要密切注意，勿使漏斗全部埋入水中，以防倒吸。

2. 取一支试管加入 0.5 mL 水，再收集一滴馏出液，检验其是否溶于水。

3. 浓硫酸的作用是溶解并除去粗品中少量未反应的正丁醇和副产物(1-丁烯、2-丁烯和丁醚等杂质)。否则正丁醇和 1-溴丁烷可以形成共沸物(沸点为 98.6 ℃，其中正丁醇含量为 13%)而难以除去。

4. 在后处理过程中，经历多次洗涤，需正确判断产物层。

【操作注意事项】

使用浓硫酸时要特别小心，以防灼伤。

【废物处理】

1-溴丁烷蒸馏的残余液倒入卤化有机溶剂专门回收容器；所有水溶液用水稀释（反应残馏液、硫酸洗涤液和氢氧化钠洗涤液），再用碳酸钠混合并中和后排入下水道；湿干燥剂放在指定的废物容器中。

【思考题】

1. 请说明本实验中浓硫酸的作用。

2. 请用反应方程式说明丁醚、1-丁烯和 2-丁烯等反应的副产物是如何用硫酸除去的？

3. 为什么在蒸馏 1-溴丁烷前需用无水氯化钙干燥？为什么蒸馏前一定将它除去？

3.3　硝基苯的制备

【实验目的】

1. 通过硝基苯的制备加深对芳烃亲电取代反应的理解。

2. 掌握液体干燥、减压蒸馏和机械搅拌的实验操作方法。

【实验原理】

硝化反应是制备芳香硝基化合物的主要方法，也是重要的亲电取代反应之一。芳烃的硝化反应比较容易进行，通常在浓硫酸存在的情况下与浓硝酸反应，烃的氢原子被硝基取代，生成相应的硝基化合物。硫酸的作用是提供强酸性的介质，有利于硝酰阳离子（N^+O_2）的生成，它是真正的亲电试剂。硝化反应通常在较低的温度下进行，在较高的温度下由于硝酸的氧化作用往往导致原料的损失，其反应式如下：

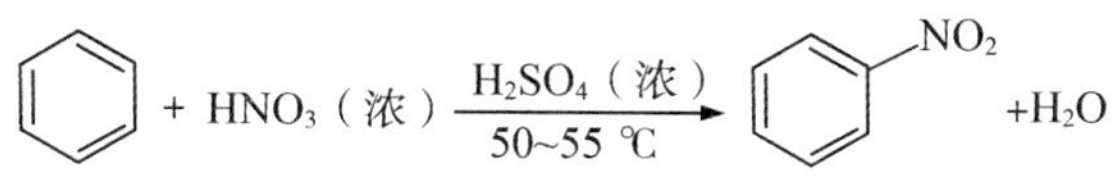

$$C_6H_6 + HNO_3\text{（浓）} \xrightarrow[50\sim55\ ℃]{H_2SO_4\text{（浓）}} C_6H_5NO_2 + H_2O$$

【仪器与试剂】

仪器：回流冷凝管，三口圆底烧瓶，恒压滴液漏斗，机械搅拌器，Y 形管，温度计，分液漏斗，减压蒸馏装置，油浴加热。

试剂：苯，浓硝酸，浓硫酸，氢氧化钠，无水氯化钙等。

【实验步骤】

1. 在 100 mL 锥形瓶中，加入 18 mL 浓硝酸，在冷却和摇荡下慢慢加入 20 mL 浓硫酸制成混合酸备用。

2. 按图 3.3.1 所示装好反应装置。

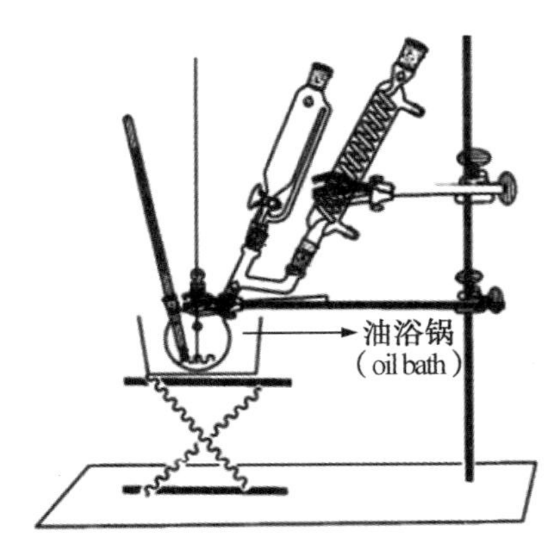

图 3.3.1 硝基苯的制备反应装置图(Apperatus for the preparation of nitrobenzene)

3. 配好的混酸放入恒压滴液漏斗中，在 250 mL 三颈圆底烧瓶中放入 18 mL 苯。

4. 启动搅拌，将混酸慢慢滴入反应瓶，控制反应温度在 50～55 ℃。

5. 在冷水浴中冷却反应混合物，然后将其移入 100 mL 分液漏斗，放出下层混合酸。有机层依次用等体积(约 20 mL)的水、5% NaOH 溶液、水洗涤后，将硝基苯移入内含 2 g 无水氯化钙的 50 mL 锥形瓶中，旋摇至混浊消失。

6. 将干燥好的硝基苯滤入 50 mL 干燥圆底烧瓶中，接空气冷凝管，加热蒸馏，收集 205～210 ℃馏分，得到硝基苯，产率为 70%～80%。

【操作注意事项】

1. 硝基化合物对人体的毒性较大，所以处理硝基化合物时要特别小心，若不慎触及皮肤，应立即用少量乙醇清洗，也可用肥皂和温水清洗。

2. 洗涤硝基苯时，加入 NaOH 后不可过分用力振荡，否则将导致产品乳化难以分层。遇此情况，可加入固体 NaOH 或 NaCl 饱和溶液，并滴加数滴酒精静置片刻即可分层。

3. 因残留在烧瓶中的硝基苯在高温时易发生剧烈分解，故蒸馏产品时不可蒸干或使温度超过 114 ℃。

4. 硝化反应是一个放热反应，实验时温度不可超过 55 ℃。

【废物处理】

易燃易爆物质的残渣(如金属钠、白磷、火柴头)不得倒入污物桶或水槽中，应收集在指定的容器内。废液，特别是强酸和强碱，不能直接倒在水槽中，应稀释后倒入水槽，再用大量自来水冲洗水槽及下水道。有毒物质应按实验室的规定办理审批手续后领取，使用时严格操作，用后妥善处理。

【思考题】

1. 本实验为什么要控制反应温度在 50～55 ℃，温度过高或过低各有什么影响?

2. 粗产物依次用水、碱液、水洗涤的目的是什么?

3.4 对硝基苯甲酸的制备

【实验目的】

1. 掌握对硝基苯甲酸制备的原理及方法。

2. 掌握固体有机化合物的纯化方法。

【实验原理】

制备芳香酸常用侧链氧化法，即用重铬酸钾（钠）或高锰酸钾、硫酸等氧化剂将苯环上的烷基氧化成羧基。侧链氧化法生成的粗产品为酸性固体物质，可通过加碱溶解、再酸化的方法来纯化，其反应如下：

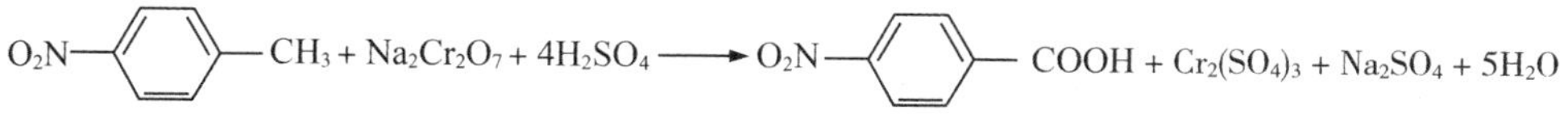

$$O_2N-C_6H_4-CH_3 + Na_2Cr_2O_7 + 4H_2SO_4 \longrightarrow O_2N-C_6H_4-COOH + Cr_2(SO_4)_3 + Na_2SO_4 + 5H_2O$$

【仪器与试剂】

仪器：三颈瓶，圆底烧瓶，回流冷凝管，电磁加热搅拌器，滴液漏斗，抽滤瓶，烧杯。

试剂：对硝基甲苯，$Na_2Cr_2O_7$，浓硫酸，15%硫酸，5% NaOH 溶液。

【实验步骤】

1. 在 100 mL 三颈瓶中安装带搅拌、回流、滴液的装置后，依次加入 3 g 对硝基甲苯、9 g 重铬酸钾粉末以及 20 mL 水。

2. 边搅拌边通过滴液漏斗向三颈瓶中滴入 12.5 mL 浓硫酸。注意：用冷水冷却，以免对硝基甲苯因温度过高挥发而凝结在冷凝管上。

3. 硫酸滴加完后，加热回流，反应液呈黑色。此过程中，冷凝管可能会有白色的对硝基甲苯析出，可适当关小冷凝水，使其熔融滴下。

4. 待反应物冷却后，边搅拌边加入 40 mL 冰水，有沉淀析出。然后，抽滤并用 25 mL 水分两次洗涤。

5. 将洗涤后的对硝基苯甲酸的黑色固体放入盛有 15 mL 5%硫酸中，沸水浴上加热 10 min，冷却后抽滤（目的是除去未反应完的铬盐）。

6. 将抽滤后的固体溶于 25 mL 5% NaOH 溶液中，50 ℃温热后抽滤，在滤液中加入 0.5 g 活性炭，煮沸趁热抽滤（此步操作很关键，温度过高对硝基甲苯融化被滤入滤液中，温度过低对硝基苯甲酸钠会析出，影响产物的纯度或产率）。

7. 充分搅拌下将抽滤得到的滤液慢慢加入盛有 30 mL 15%硫酸溶液的烧杯中析出黄色沉淀，抽滤，少量冷水洗涤两次，干燥后称重（加入顺序不能颠倒，否则会造成产品不纯）。

8. 用乙醇-水混合溶剂重结晶。注意：对硝基苯甲酸的熔点为 241～242 ℃。

【废物处理】

酸碱废液的处理：实验室中各类酸碱溶液的用量较大，用后需按照酸碱性质进行分类收集，并用中和法进行处理。在查明两种废液可相互混合的情况下，可分次少量将其相互中和，然后用水稀释达标后排放。

含氧化剂和还原剂废液的处理：这类废液的处理常采用氧化还原法，首先是将含氧化剂、还原剂的废液分别收集，查明两者混合没有危险后，将其中一种废液分次少量加入到另一种废液中进行无害化处理。

【操作注意事项】

1. 从滴加浓硫酸开始，整个反应过程中，要一直保持搅拌。

2. 滴加浓硫酸时，只搅拌，不加热。加浓硫酸的速度不能太快，否则会引起剧烈反应。

3. 转入 40 mL 冷水中后，可用少量(约 10 mL)冷水洗涤烧瓶。

4. 用碱溶解时，可适当温热，但温度不能超过 50 ℃，以防未反应的对硝基甲苯熔化，进入溶液。

5. 酸化时，应将滤液倒入酸中，不能反过来将酸倒入滤液中。

6. 纯化后的产品，用蒸汽浴干燥。

【思考题】

1. 芳基侧链的氧化方法有哪些？氧化规律有哪些？

2. 提高非均相反应的措施除电动搅拌外，还有哪些措施？

3. 为什么酸化时，要将滤液倒入酸中，而不能反过来将酸倒入滤液中？

3.5 己二酸的绿色合成

【实验目的】

1. 通过己二酸的制备，了解传统合成方法的劣势和绿色合成方法的优势。

2. 熟悉催化剂无需回收条件下的循环使用。

【实验原理】

己二酸俗称“肥酸”，分子式为 $C_6H_{10}O_4$。对于己二酸的生产，目前全世界应用最广泛的是以环己乙醇或环己酮为原料的硝酸氧化工艺路线。在此生产路线中，使用强氧化性的硝酸会严重腐蚀设备，而且生产过程中产生的 N_2O 气体被认为是引起全球变暖和臭氧减少的原因之一，会给环境造成极大的污染。日本科学家野依良治(Ryoji Noyori)于 1998 年在 *Science* 上发表了一篇有关己二酸的绿色合成方法，提出了用水作溶剂，过氧化氢(H_2O_2)作氧化剂，钨酸钠(Na_2WO_4)作催化剂，在硫酸氢钾($KHSO_4$)的参与下，甲基三辛氯化铵(Aliquat 336)作相转移催化剂的绿色制备路线。此路线不用强酸，不产生 N_2O 有害气体，用水作溶剂不产生废液，并且催化剂不用回收可以直接循环使用。此后，又有研究人员发现无须相转移催化剂，采用水作溶剂，H_2O_2 作催化剂，利用钨酸钠-草酸原位合成的配位催化剂亦可催化合成己二酸，其反应式如下：

$$\text{环己烯} \xrightarrow[\text{KHSO}_4\text{, Aliquat 336}]{\text{NaWO}_4 \cdot 2\text{H}_2\text{O} \cdot \text{H}_2\text{O}_2} \text{HOOC(CH}_2)_2\text{COOH}$$

本实验以环己烯为原料，在钨酸钠-草酸催化下，双氧水氧化合成乙二酸，其反应式如下：

$$\text{环己烯} \xrightarrow[\text{H}_2\text{C}_2\text{O}_4 \cdot 2\text{H}_2\text{O} \cdot \text{H}_2\text{O}_2]{\text{NaWO}_4 \cdot 2\text{H}_2\text{O}} \text{HOOC(CH}_2)_2\text{COOH}$$

【仪器与试剂】

仪器：磁力加热搅拌器，冷凝管，圆底烧瓶，烧杯，干燥管，表面皿，碱式滴定管，熔点测定仪，三颈烧瓶，真空泵，分析天秤。

试剂：钨酸钠（AR），环己醇（AR），环己烯（AR），浓硝酸（AR），硫酸氢钾（AR），甲基三辛基氯化铵（AR），过氧化氢（AR），草酸（AR）。

【实验步骤】

1. 传统制备方法

将 2 mL 浓硝酸加入到 10 mL 圆底烧瓶中，放入搅拌磁子，装上冷凝管，装置图如图 3.5.1 所示。在通风橱里（因为有 N_2O 气体放出）加热到 80 ℃。接着将 1 mL 环己醇通过滴管从冷凝管上方缓慢滴加到圆底烧瓶中，在滴加的过程中，尽量避免环己醇与冷凝管内壁接触，滴加速度控制在 1 滴/min 左右，滴加过程持续 30～40 min。滴加完后，保持 80 ℃继续反应 2 h，然后冷却至室温。放入冰水浴中冷却，晶体析出。用布氏漏斗抽滤，先用少量冰水洗涤，再用少量乙醇洗涤，尽量抽干晶体。收集药品在干燥箱中进行干燥15 min，称量计算产率。最后，取少量产品测定熔点，纯己二酸的熔点为 152 ℃。

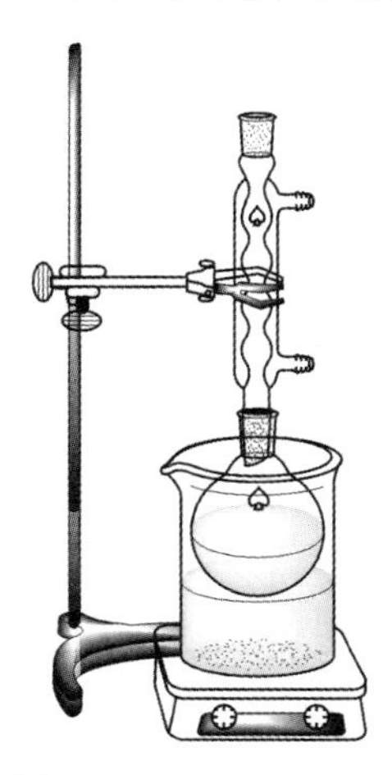

图 3.5.1 传统制备方法装置图（Device diagram of traditional preparation metho）

2. 绿色制备方法(钨酸钠-草酸作催化剂)

在 100 mL 三颈烧瓶中加入 1.50 g 钨酸钠，0.57 g 草酸，34.0 mL 30%的 H_2O_2 和搅拌磁子。室温下搅拌 15～20 min 后，加入 6.00 g 环己烯。搭好回流装置(见图 3.5.2)，继续快速剧烈搅拌并加热至回流。反应过程中回流温度将慢慢升高，回流 1.5 h 后，将热的反应液倒入 100 mL 烧杯中，冷水冷却至接近室温后，再在冰水浴中冷却 20 min。用布氏漏斗抽滤，先 5 mL 冰水洗涤后，再用 5 mL 乙醇洗涤，尽量抽干晶体，将产品干燥至恒重后称量，记录己二酸的重量，计算产率(约为 30%)，取少量产品测定熔点。

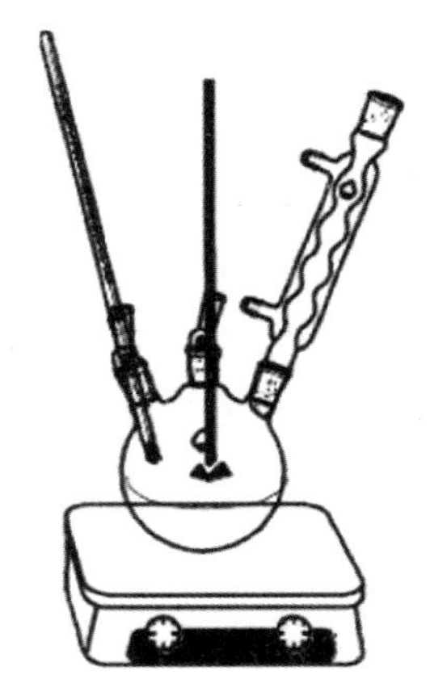

图 3.5.2 绿色制备方法装置图(device diagram of green preparation method)

3. 己二酸含量的测定

用酸碱滴定法测定己二酸的含量：分别准确称取两次实验产品 0.1 g(准确至 0.0001 g)于 250 mL 锥形瓶中，加入 50 mL 热的蒸馏水，搅拌溶解样品。加入 1 滴酚酞指示剂，用 0.1 mol/L 的氢氧化钠标准溶液滴定至微红色，30 s 内不退色即为终点。各滴定 2 份，计算样品中己二酸的含量。

【结果记录】

实验完成后将结果填入表 3.5.1。

表 3.5.1 结果记录表

制备方法	原料质量/g	理论产量/g	实际产量/g	产率/%	熔点/℃	纯度/%
传统方法						
绿色方法						

【操作注意事项】

1. 此反应为放热反应，反应开始后会使混合物超过 45 ℃，假如在室温下开始反应 5 min 后，混合物温度还不能上升至 45 ℃，则可小心加热至 40 ℃，使反应开始。

2. 要不断震荡或搅拌，否则反应物极易暴沸冲出容器。

3. 为了提高产率，最好用冰水冷却溶液以降低己二酸在水中的溶解度。

【废物处理】

处理含甲醇、乙醇和醋酸等废液的溶液时，可用蒸馏法精制后回收利用。如果甲醇、乙醇和醋酸浓度较低，可以用大量的水稀释达标后排放。

【实验思考】

本实验过程中，若加入原料过早，会导致反应提前进行，最终导致产率较低，且熔程较长。

3.6 肉桂酸的制备

【实验目的】

1. 了解肉桂酸的制备原理和方法。

2. 掌握回流、水蒸气蒸馏等基本操作。

【实验原理】

利用柏琴(Perkin)反应，将苯甲醛与酸酐混合后在相应的羧酸钾或钠盐催化下加热，可以制得 α,β-不饱和酸，其中碱的作用是促使酸酐烯醇化。本实验用碳酸钾代替醋酸钾，反应时间短，产率高，其反应式如下：

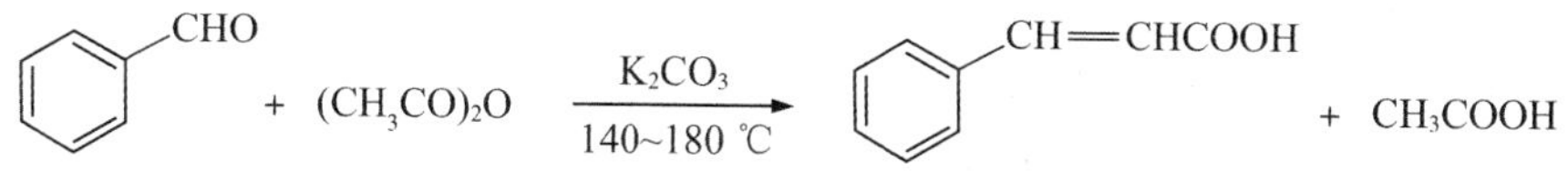

$$C_6H_5CHO + (CH_3CO)_2O \xrightarrow[140\sim180\ ℃]{K_2CO_3} C_6H_5CH{=}CHCOOH + CH_3COOH$$

【仪器与试剂】

仪器：圆底烧瓶，冷凝管，蒸馏接头，接收管，加热套，布氏漏斗，烧杯，量筒，抽滤瓶，真空水泵，滤纸。

试剂：苯甲醛，乙酸酐，无水碳酸钾，10%氢氧化钠溶液，浓盐酸，刚果红试纸。

【实验步骤】

在 250 mL 圆底烧瓶中放入 3 mL 新蒸过的苯甲醛，8 mL(0.085 mol)新蒸过的乙酸酐以及研细的 4.2 g 无水碳酸钾，用加热套加热回流 45 min。

待反应液冷却后，加入 20 mL 水，将瓶内生成的固体尽量捣碎(捣碎过程中要小心)，用水蒸气蒸馏法蒸出未反应的苯甲醛。

再将烧瓶冷却，加入 20 mL 10%氢氧化钠溶液，以保证所有的肉桂酸成钠盐而溶解。抽滤，将滤液倾入 250 mL 烧杯中，冷却至室温，边搅拌边加入浓盐酸，酸化至刚果红试纸变蓝(pH 值为 2～3)。

冷却，抽滤，用少量水洗涤沉淀，抽干，干燥，称重，计算产率。如需精制，可用热水或

3∶1的稀乙醇重结晶。

纯肉桂酸(反式)为无色晶体,熔点为133～136 ℃。

【解释说明】

1. 苯甲醛久置,容易自动氧化成苯甲酸,既影响反应的进行,也不易除干净,将影响产品的质量。故所需的苯甲醛要事先蒸馏,收集170～180 ℃馏分供使用。

2. 乙酸酐久置会因吸潮和水解转变为乙酸,故本实验的醋酐必须在实验前重新蒸馏。

3. 由于实验有二氧化碳放出,故反应初期有泡沫产生。

4. 肉桂酸有顺、反异构体,通常制得的是其反式异构体。

【废物处理】

所有滤液应倒入指定的回收容器中进行回收。

【思考题】

1. 用水蒸气蒸馏除去什么?为什么能用水蒸气蒸馏纯化产品?

2. 苯甲醛与丙酸酐在无水碳酸钾存在下的反应产物是什么?

3. 酸化时,能否使用硫酸?

3.7 乙酸乙酯的制备

【实验目的】

1. 了解酯化反应的原理和方法。

2. 学习回流操作,巩固蒸馏及萃取分离技术。

【实验原理】

酯化反应是指酸与醇作用生成酯和水的过程,其逆反应为酯的水解反应。酯化反应一般采用强酸作催化剂,如浓硫酸、干燥氯化氢、有机强酸和阳离子交换树脂等,其中浓硫酸最常用,加入强酸的目的是使反应迅速达到平衡。酯化反应式如下:

$$R-\overset{\overset{O}{\|}}{C}-OH + R'OH \xrightleftharpoons{H_2SO_4} R-\overset{\overset{O}{\|}}{C}-OR + H_2O$$

为了提高反应的收率,往往采用醇或酸过量,或者除去反应生成的酯和水,以使平衡向右移动。

本实验采用乙醇过量的方法来提高反应收率。乙酸乙酯和水形成的共沸混合物(70.4 ℃)比乙醇(78 ℃)和乙酸(118 ℃)的沸点都低,而乙酸乙酯的沸点为77.06 ℃。因此,乙酸乙酯很容易蒸出。

除生成乙酸乙酯外,本实验还存在生成乙醚的副反应。乙酸乙酯的制备反应式如下:

$$CH_3COOH + CH_3CH_2OH \xrightleftharpoons{H_2SO_4} CH_3COOC_2H_5 + H_2O$$

【仪器与试剂】

仪器:50 mL、100 mL 圆底烧瓶,100 mL 分液漏斗,球形冷凝管,50 mL 三角瓶,25 mL 量筒,蒸馏装置。

试剂:无水乙醇,冰乙酸,浓硫酸,饱和 NaCl 溶液,饱和 $CaCl_2$ 溶液,饱和 Na_2CO_3 溶液,无水 Na_2SO_4。

【实验步骤】

在 100 mL 圆底烧瓶中放入 20 mL 无水乙醇(15.7 g, 0.34 mol)、12 mL 冰乙酸(12.59 g, 0.21 mol),在振摇下缓慢加入 5 mL 浓硫酸,混匀后加入 2～3 粒沸石,装上球形冷凝管,水浴加热回流半小时。将反应液冷却后,加入 2～3 粒沸石,改成蒸馏装置。水浴蒸馏至无液体流出时为止,得到粗乙酸乙酯。

向粗乙酸乙酯中加入饱和的 Na_2CO_3 水溶液,直至有机层(上层)呈中性。将溶液转入 100 mL 分液漏斗中,静置分层,除去下层水溶液,酯层用等体积饱和 NaCl 溶液洗涤后,再用等体积饱和 $CaCl_2$ 水溶液洗涤两次,除去下层水溶液,转移到干燥的三角瓶中,用无水 $MgSO_4$ 干燥。

将干燥过的乙酸乙酯滤入 50 mL 干燥的圆底烧瓶内,加入 2～3 粒沸石,在水浴上加热蒸馏,收集 75～80 ℃馏分。纯乙酸乙酯为无色液体,沸点为 77.06 ℃,比重为 0.901,稍溶于水,具有水果香。本实验的产率为 60%～67%。

【解释说明】

1. 当温度超过 120 ℃时,会增大乙醚的生成量,故回流时保持微沸即可。

2. 馏出液中除酯和水外,还含有少量的乙醇和醋酸。用碳酸钠除去醋酸、亚硫酸(硫酸被还原产生),用氯化钙溶液除去乙醇。但碳酸钠用量不宜过多,否则下一步用氯化钙处理时会形成乳胶。

【废物处理】

普通的有机废液,如石油醚、乙酸乙酯等,可直接倒入废液桶中,废液桶尽量不要密封,不能装太满。有刺激性气味的液体倒入另一个废液桶内,立即封盖,统一处理。

【思考题】

1. 本实验中浓硫酸起什么作用?

2. 蒸出的粗乙酸乙酯中有哪些杂质?

【知识拓展】

香精香料酯

3.8 乙酰水杨酸的制备

乙酰水杨酸的制备(实验原理)

乙酰水杨酸的制备(操作视频)

【实验目的】

学习乙酰水杨酸的制备方法及制备原理

【实验原理】

乙酰水杨酸通常称为阿司匹林(aspirin),是由水杨酸(邻羟基苯甲酸)和乙酸酐合成的。直到目前,阿司匹林仍然是一个被广泛使用的,具有解热、镇痛、抗炎作用的药物。水杨酸是一个具有酚羟基和羧基的双官能团化合物,能进行两种不同的酯化反应。当水杨酸与乙酸酐反应时,可以得到乙酰水杨酸,即阿司匹林。当水杨酸与过量的甲醇反应时,生成水杨酸甲酯。本实验将进行前一个反应的实验,其反应式如下:

$$o\text{-}HOC_6H_4COOH + (CH_3CO)_2O \xrightleftharpoons[80\sim85\ ℃]{H_3PO_4} o\text{-}CH_3COOC_6H_4COOH + CH_3COOH$$

在实验过程中,生成乙酰水杨酸的同时,水杨酸分子之间发生缩合反应,生成少量聚合物,其反应式如下:

$$o\text{-}HOC_6H_4COOH \xrightarrow{H^+} \left[-O-C_6H_4-C(=O)-O-C_6H_4-C(=O)-O-C_6H_4-C(=O)-O- \right]_n + H_2O$$

乙酰水杨酸能与碳酸氢钠反应生成水溶性钠盐,而副产物的聚合物不能溶于碳酸氢钠,这种性质差别可用于阿司匹林的纯化。

可能存在于最终产物中的杂质是水杨酸本身,这是由于乙酰化反应不完全或由于产物在分离步骤中发生水解造成的,它可以在各步纯化过程和产物重结晶中被除去。与大多数酚类化合物一样,水杨酸可与三氯化铁形成深色络合物,因乙酰水杨酸的酚羟基已被酰化,不再与三氯化铁发生颜色反应,因此杂质很容易被检出。

【仪器与试剂】

仪器:SHZ-C型循环水式多用真空泵(公用),电热恒温水浴锅,干燥箱,架盘药物天平,不锈钢刮铲,剪子,50 mL锥形瓶,温度计(0~100 ℃),10 mL量杯,50 mL烧杯,抽滤瓶,布氏漏斗,定性滤纸,称量纸,药匙,试管架及试管,回收瓶,木夹子,玻璃棒,球形冷凝管，50 mL 圆底烧瓶。

试剂：水杨酸，乙酸酐，磷酸，95%乙醇，0.1%三氯化铁溶液，浓盐酸，饱和碳酸氢钠溶液。

【实验步骤】

在干燥的 50 mL 烧瓶中加入 3 g(0.022 mol)干燥的水杨酸，再缓缓加入 4.5 g(0.04 mol)乙酸酐，摇匀后，滴加 5 滴磷酸，充分摇动，使水杨酸全部溶解。将烧瓶和冷凝管放入 80～85℃水浴中，恒温 10～15 min，期间不断振摇。稍冷，在不断搅拌下倒入50 mL 水，并用冷水冷却，抽滤，用适量冷水洗涤。将抽滤后的粗产物转入 100 mL 烧杯中，在搅拌下加入 38 mL 饱和碳酸氢钠水溶液。加完后继续搅拌几分钟，直至无二氧化碳产生。抽滤，滤出副产聚合物，并用 5～10 mL 水冲洗漏斗，合并滤液，倒入预先盛有7 mL 浓盐酸和 15 mL 水的烧杯中，搅拌均匀，即有乙酰水杨酸晶体析出。将烧杯用冰水冷却，使晶体结晶完全。最后，抽滤，并用冷水洗涤结晶。将结晶转移至表面皿，干燥后称重，计算产率。取几粒结晶加入盛有 1 mL 95%乙醇的试管中溶解，加入 1～2 滴 0.1 三氯化铁溶液，观察有无颜色变化，从而判断产物中有无未反应的水杨酸。本实验的产率约为 60%。

【解释说明】

1. 乙酸酐应是新制的。

2. 乙酰水杨酸受热易分解，因此熔点不明显，它的分解点为 128～135 ℃。测定熔点时，应先将载体加热至 120 ℃，然后加入样品测定。

【操作注意事项】

1. 仪器要全部干燥，药品要进行干燥处理，乙酰酐应为新制的，否则产率会很低。水浴加热温度不宜过高，时间不宜过长，否则副产物可能增加。

2. 若不结晶，可用玻璃棒摩擦瓶壁或置于冰水中冷却。

3. 重结晶加热时间不宜过长，并需控制水温，且产物应自然晾干

【废物处理】

实验完毕后，产物回收，将仪器洗涤干净，废液回收到废液桶中。

【思考题】

1. 反应仪器为何要求干燥？

2. 酰化反应加入磷酸的目的是什么？

3. 在制备乙酰水杨酸的过程中，你认为有哪些问题需要引起关注？如何确保有较高的产率？

4. 乙酰水杨酰粗产品能否在热水中重结晶？

【知识拓展】

镇痛药

3.9 乙酰苯胺的制备

【实验目的】

1. 熟悉苯胺酰化的原理和操作步骤。

2. 掌握重结晶的操作方法。

【实验原理】

苯胺很容易进行酰化反应,常用的酰化试剂有冰醋酸、乙酸酐和乙酰氯等。乙酰苯胺是有机合成的一个重要中间体,也是合成磺胺类药物的重要原料,它又被称为"退热冰",是一种强退烧药,但具有较大的毒副作用。胺的乙酰化反应在有机合成中也非常重要,具有保护芳香环上的氨基,使其不被反应试剂破坏的作用。氨基经乙酰化保护后,尽管其定位效应不改变,但对芳环的活化能力降低了,因而使反应的多元取代变为一元取代。由于空间效应,往往对位的反应活性比邻位高,易生成选择性的对位取代物。本实验用冰醋酸作乙酰化试剂,不仅经济,且符合"绿色化学"的要求,其反应式如下:

$$C_6H_5NH_2 + CH_3COOH \xrightarrow{Zn} C_6H_5NHCOCH_3$$

【仪器与试剂】

仪器:圆底烧瓶,分馏柱,蒸馏接头,冷凝管,加热套,温度计(0～150 ℃),滤纸,烧杯,真空水泵,量筒,抽滤瓶,布氏漏斗。

试剂:新蒸馏苯胺,冰醋酸,锌粉。

【实验步骤】

在 50 mL 圆底烧瓶中倒入 10 mL(10.2 g,0.11 mol)苯胺,加入 15 mL(15.7 g,0.26 mol)冰醋酸和 0.1 g 锌粉,在圆底烧瓶上装上一个分馏柱,插上温度计,用一个 50 mL 的圆底烧瓶作接收器,装置如图 3.9.1 所示。

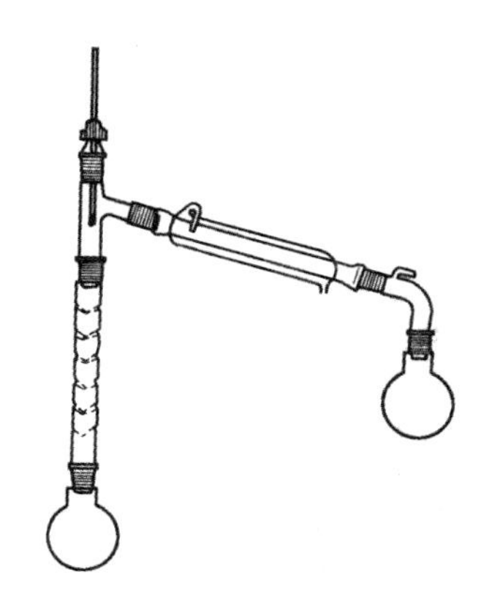

图 3.9.1 乙酰苯胺的制备装置图(reaction device)

将圆底烧瓶缓缓加热,至反应混合物保持微沸约 15 min,然后控制加热速度,保持蒸馏的气温在 105 ℃左右,反应生成的水及少量冰醋酸会被蒸出。反应约 1.5 h 后,温度计读数下降,表示反应已经完成。边搅拌边趁热将反应物倒入盛有 200 mL 冰水的烧杯中,乙酰苯胺生成。抽滤,用冰水洗涤,并用水重结晶,计算产率,测定乙酰苯胺的熔点。

纯的乙酰苯胺为无色片状晶体,熔点为 113～114 ℃。

【解释说明】

1. 苯胺有毒,它能经过皮肤被人体吸收,使用时需小心。久置的苯胺因被氧化而色深有杂质,最好用新蒸的苯胺。

2. 加入少量锌粉的目的是防止苯胺在反应过程中被氧化,如果加入过多,会形成不溶于水的氢氧化锌。

3. 控制温度是为了尽量除去反应中生成的水,防止冰醋酸被蒸出。

【废物处理】

过滤、重结晶等操作的水溶液应倒入指定的废水容器中。

【思考题】

1. 反应时为什么要控制冷凝管上端的温度在 105 ℃左右?

2. 为什么洗涤产品时要用冰水?

3. 用苯胺作原料进行苯环上的某些取代反应时,为什么常常要先进行酰化?

第 4 章　综合性和设计性实验

4.1　从茶叶中提取咖啡碱

从茶叶中提取咖啡碱(实验原理)

从茶叶中提取咖啡碱(操作视频)

【实验目的】

1. 通过实验了解从植物中提取生物碱的一般方法。

2. 掌握索氏提取器的使用方法,学习用升华法或溶剂萃取法提纯有机化合物的方法。

3. 了解生物碱的一般鉴定方法。

【实验原理】

植物中的生物碱常以盐(能溶于水或醇)或游离碱(能溶于有机溶剂)的状态存在。因此,可用水、醇或其他有机溶剂提取生物碱。将生物碱与提取液中其他杂质分离时,可根据生物碱与这些杂质在溶剂中的溶解度不同、化学性质不同来具体对待。

茶叶中含有的生物碱均为黄嘌呤衍生物,有咖啡碱、茶碱、可可碱等,其中以咖啡碱含量最多,为 2%～5%。咖啡碱为无臭、味苦的白色结晶,熔点为 235～236 ℃,可溶于水,易溶于热水、氯仿。提取咖啡碱的方法主要有连续萃取法和浸取法。

连续萃取法以乙醇为溶剂,在索氏提取器中连续抽提,然后蒸出溶剂得到粗咖啡因。粗咖啡因中还含有其他生物碱和杂质(如丹宁酸等),可利用升华法进一步提纯。升华法制得的产品通常纯度较高,但损失也较大。含结晶水的咖啡因加热至 100 ℃时失去结晶水开始升华,120 ℃时显著升华,176 ℃时迅速升华。

浸取法利用咖啡碱易溶于热水的性质可将其自茶叶中提出。茶叶中的鞣酸亦溶于水,因此会有 11%～12%的鞣酸会随咖啡碱一同被提出,可利用鞣酸与醋酸铅生成沉淀的性质将鞣酸除去,然后再利用咖啡碱溶于乙酸乙酯的性质使其与其他水溶性杂质分离。

浸取法提取咖啡碱的流程图如图 4.1.1 所示。所得咖啡碱可以通过紫脲酸铵反应或碘化铋钾试剂加以鉴别。

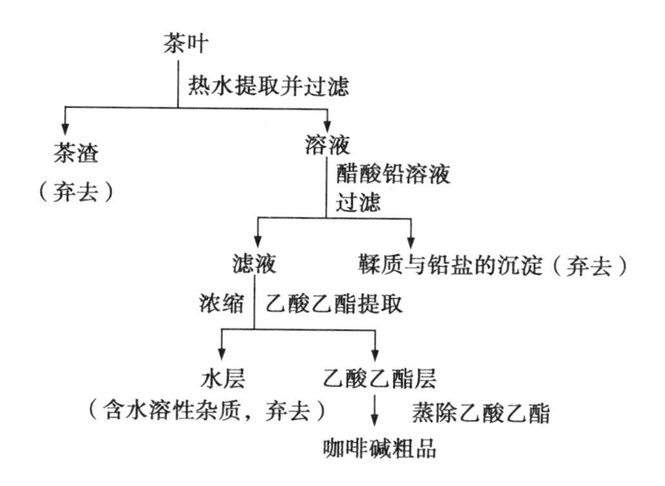

图 4.1.1　浸取法提取咖啡碱的流程图(The flow chart of caffeine by leaching)

【仪器与试剂】

仪器：索氏提取器，常压蒸馏装置，抽滤装置，恒温水浴锅，电炉，分液漏斗，长颈漏斗，蒸发皿，坩埚，铁架台，石棉网，烧杯，量筒，试管，滤纸，脱脂棉，沸石。

试剂：茶叶(市售)，10%醋酸铅溶液，乙酸乙酯，饱和食盐水，氯酸钾，浓盐酸，浓氨水，碘化铋钾试剂，5%硫酸溶液，95%乙醇，生石灰粉。

【实验步骤】

1. 咖啡碱的分离

连续萃取法：

(1)称取 8 g 茶叶，装入索氏提取器的滤纸筒内，在提取器的烧瓶中加入 80 mL 95%的乙醇和几块沸石，安装索氏提取器(见图 4.1.2)，接通冷凝水，加热回流，连续抽提 1～1.5 h(当提取液颜色很淡时即可停止抽取)，待冷凝液刚刚虹吸下去时，立即停止加热，冷却。

(2)装好蒸馏装置，加热蒸馏，回收大部分乙醇。把残液(15～20 mL)倒入蒸发皿中，蒸馏瓶用很少量酒精洗涤，洗涤液合并于蒸发皿中，在电热套上浓缩至残液约 10 mL 左右。

(3)在盛有浓缩残液的蒸发皿中加入 4 g 生石灰(CaO)粉，搅拌均匀。然后将蒸发皿固定到铁环上，采用电热套加热，不断搅拌，蒸干成松散状。压碎块状物，小火力焙炒，除尽水分。冷却后，擦去沾在蒸发皿边沿上的粉末，以免升华时污染产物。

(4)在上述蒸发皿上盖一张刺有许多小孔的圆滤纸，在上面罩上干燥的玻璃漏斗，漏斗颈部塞少许脱脂棉，以减少咖啡因蒸气逸出，升华装置如图 4.1.3 所示。使用电热套加热蒸发皿进行升华，当滤纸上出现白色针状结晶时，停止加热。冷却(约 5 min)后小心地揭开漏斗和滤纸，仔细地把附在滤纸及器皿周围的咖啡因晶体(白色、针状)用小刀刮入干燥、洁净、已称重的 50mL 的烧杯中。残渣经拌和后，用较大火力再继续加热升华一次(或两次)。合并各次升华收集的咖啡因结晶，称重。

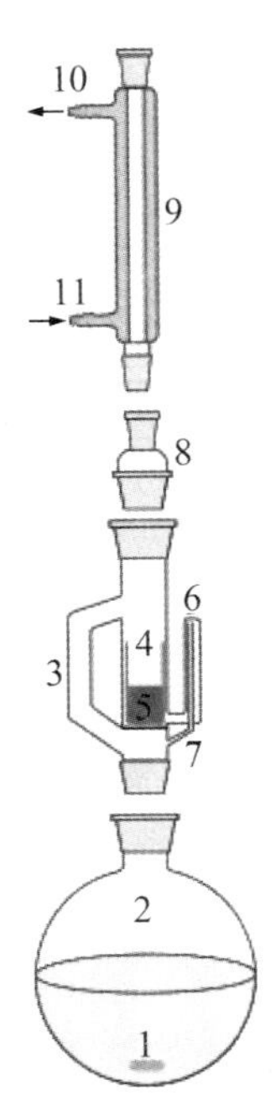

1—搅拌子(stirrer);2—烧瓶;3—蒸汽路径(steam path);4—套管(sleeve);5—固体(solid);6—虹吸管(siphon tube);7—虹吸出口(siphon outlet); 8—转接头(Adapter); 9—冷凝管(condenser); 10—冷却水出口(cooling water outlet); 11—冷却水入口(cooling water inlet)

图 4.1.2 索氏提取器的示意图(Schematic diagram of Soxhlet extractor)

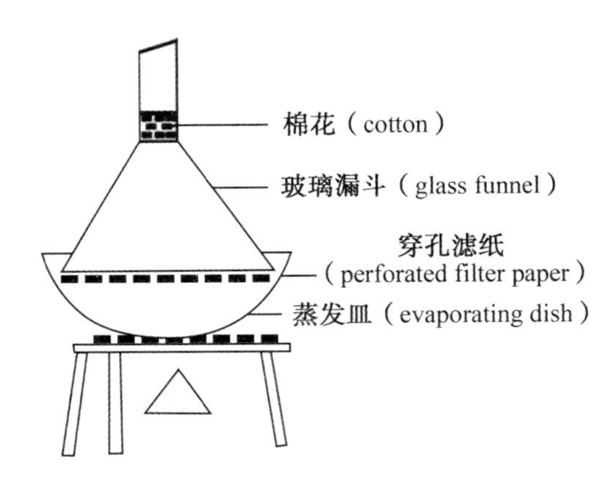

图 4.1.3 升华装置示意图(Schematic diagram of sublimation device)

浸取法：

(1)取 200 mL 烧杯 1 只，加入 5 g 茶叶及 100 mL 热水，加热煮沸约 15 min(若水分蒸发太多，可加一个表面皿)，用脱脂棉过滤除去茶渣，边搅动边向热的滤液中逐滴加入 10%醋酸铅溶液 15～20 mL，直至不再有沉淀生成。

(2)将上述混悬液加热 5 min 后抽滤，放冷。若此时又有沉淀析出，可再行减压过滤除去。

(3)将上述液体移入分液漏斗，加入 25 mL 乙酸乙酯及 15 mL 饱和食盐水，剧烈振摇。注意：乙酸乙酯易挥发，在振摇时应常将活塞打开以使过量蒸气逸出。放置片刻，待液体分层后将上层乙酸乙酯液分出，倒入小蒸馏瓶中，加入 2～3 粒沸石，在水浴上蒸馏回收乙酸乙酯，即得咖啡碱粗品。

2. 咖啡碱的鉴定

(1)在小坩埚内加入咖啡碱粗品的结晶数粒，再加入少许氯酸钾结晶及 2～3 滴浓盐酸，然后在石棉网上加热至液体完全蒸发，冷却后加入 1 滴浓氨水，溶液呈紫色。这就是紫脲酸铵反应，此反应阳性表明有生物碱存在。

(2)在剩下的咖啡碱结晶中加入 2 mL 5％硫酸溶液，搅拌使其溶解，取约 1 mL 咖啡碱硫酸溶液于试管中，加 2 滴碘化铋钾试剂，若生成红棕色沉淀，表明生物碱存在。

【解释说明】

1. 升华是将具有较高蒸气压的固体物质加热到熔点以下，不经过熔融状态就直接变成蒸气，蒸气变冷后，又直接变为固体的过程。升华是精制某些固体化合物的方法之一，能用升华方法精制的物质必须满足以下条件：①被精制的固体要有较高的蒸气压；②杂质的蒸气压应与被纯化的固体化合物的蒸气压有显著差异。

2. 茶叶中含有的生物碱均为黄嘌呤衍生物，其结构式如下：

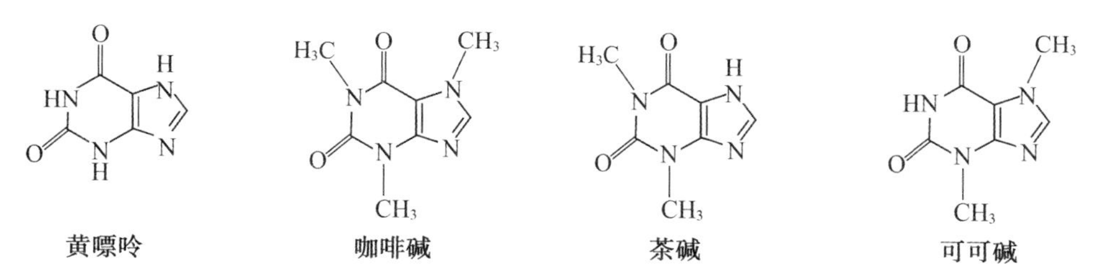

此类生物碱都具有嘌呤类的紫脲酸铵反应，其反应式如下：

(紫罗兰色)

四甲基紫酸铵

3. 滤纸筒既要紧贴器壁，又要方便取放。被提取物高度不能超过虹吸管，否则被提取物不能被溶剂充分浸泡，影响提取效果。被提取物亦不能漏出滤纸筒，以免堵塞虹吸管。

4. 生石灰(CaO)粉起到吸水和中和的作用，以除去杂质。

5. 乙酸乙酯为无色挥发性液体，有强烈的醚样气味，微带果香，易扩散，不持久，微溶于水，可与石油醚、二氯甲烷、乙醇等有机溶剂以任意比例混溶。乙酸乙酯的沸点为77 ℃，比重为0.894～0.898。

【操作注意事项】

1. 使用索氏提取器过程中，用滤纸包茶叶时要严实，防止茶叶漏出堵塞虹吸管。

2. 滤纸套筒大小要合适，既能紧贴套管内壁，又能方便取放，且其高度不能超过溶液蒸气上升装置支管口的高度，又不能低于虹吸管的最高处。

3. 烧瓶中的液体不能装得太多，一般是索氏提取器容积的3～4倍。

4. 在萃取回流充分的情况下，升华操作的好坏是本实验成败的关键，在升华过程中温度不宜过高，否则会使滤纸碳化变黑，并把一些有色物质烘出来，使产品不纯。第二次升华时，火力亦不能太大，否则会使被烘物质大量冒烟，导致产物损失。

【废物处理】

提取的咖啡碱最后进行回收，实验过程中产生的有机废液不得倒入下水道，应倒入专门的废液处理容器内，无机酸类废液也应倒入专门的集中处理容器内。

【思考题】

1. 连续萃取中生石灰起什么作用？浸取法中加入醋酸铅溶液的目的是什么？

2. 抽滤与普通过滤有何不同，其特点是什么？

3. 咖啡因的结构中，哪个氮的碱性最强？请解释原因。

4. 升华法有哪些优缺点？

【知识拓展】

咖啡因

4.2 从黄连中提取黄连素

【实验目的】

1. 学习从黄连中提取黄连素的原理和方法。

2. 进一步掌握索氏提取器的使用方法。

【实验原理】

黄连素又称小檗碱，是从黄连等中草药中分离的一种生物碱，对急性结膜炎、口疮、急性细菌性痢疾、急性肠胃炎等有很好的疗效。黄连中黄连素的含量为4%～10%，除此之外，三颗针、伏牛花、白屈菜、南天竹等中草药中也含有黄连素，但黄连和黄柏中含量最高。图4.2.1为黄连素的互变异构体，从左至右分别为季铵型、醇胺型和醛型，自然界中黄连素多以季铵碱的形式存在。

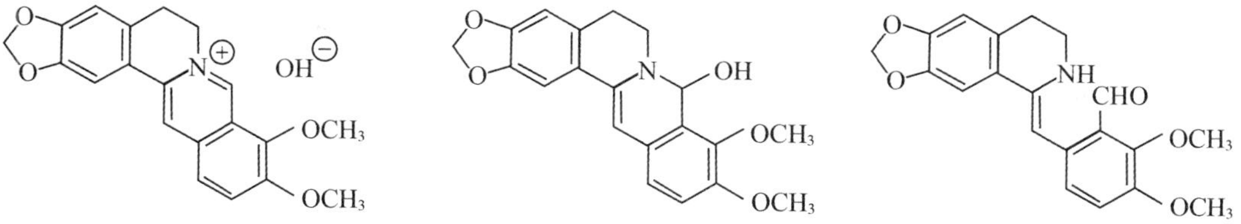

(a) **季铵型** (quaternary ammonium type)　(b) **醇胺型** (alchol amine type)　(c) **醛型** (aldehyde type)

图4.2.1　黄连素的互变异构体(Tautomers of berberine)

黄连素是黄色针状结晶，含有5.5个分子结晶水，在100 ℃干燥后，失去分子结晶水而转为棕红色。黄连素微溶于水和乙醇，较易溶于热水和热乙醇中，几乎不溶于乙醚。黄连素的盐酸盐、氢碘酸盐、硫酸盐、硝酸盐等均难溶于冷水，易溶于热水，故可利用水对其进行重结晶，从而达到纯化目的。

【仪器与试剂】

仪器：索氏提取器，减压蒸馏装置，研钵，恒温水浴锅，电炉，烘箱，布氏漏斗，抽滤瓶，分液漏斗，烧杯，量筒，滤纸，脱脂棉，沸石。

试剂：黄连，1%的醋酸溶液，丙酮(AR)，饱和石灰水，浓盐酸(AR)，0.5%硫酸溶液，95%乙醇，冰水。

【实验步骤】

1. 黄连素的提取

(1)称取10 g黄连，捣碎研磨后装入索氏提取器的滤纸套筒内，在提取器的烧瓶内加入100 mL 95%乙醇和几块沸石，安装索氏提取器提取装置(见图4.1.2)，接通冷凝水，加热回流，连续抽提1～1.5 h(当提取液颜色很淡时即可停止抽取)；待冷凝液刚刚虹吸下去时，立即停止加热，冷却。

(2)减压蒸馏，回收大部分乙醇，直至瓶内残留液体呈棕红色糖浆状，停止蒸馏。

(3)向浓缩后的有机相里加入30 mL 1%醋酸溶液，加热溶解后，趁热抽滤去掉固体杂质。向滤液中滴加浓盐酸，直至溶液浑浊为止(约需10 mL)。

(4)将上述溶液放在冰水浴中冷却，降至室温以下后有黄色针状晶体析出，抽滤，所得结晶固体用冰水洗涤两次，用丙酮洗涤一次，即可得到黄连素盐酸盐的粗产品。

(5)将粗产品放入100 mL的烧杯中，加热水至刚好溶解煮沸，用石灰乳调节pH值至8.5～9.8。冷却，滤除杂质，继续冷却至室温以下，有黄连素晶体析出。抽滤，得到黄色黄连素结晶，在50～60 ℃的烘箱中烘干，称量。

2. 黄连素的鉴定

(1)取黄连素成品少许，加浓硫酸2 mL，溶解后加入几滴浓硝酸，若溶液呈樱红色，表明黄连素存在。

(2)取黄连素成品约50 mg，加蒸馏水5 mL，缓慢加热，溶解后加入20%氢氧化钠溶液2滴，显橙色，冷却后过滤，滤液中加4滴丙酮，溶液变浑浊，放置后生成黄色的丙酮黄连素沉淀，表明黄连素存在。

(3)取黄连素少许，溶解在水中，滴加浓硝酸数滴，若产生黄绿色硝酸黄连素沉淀，表明黄连素存在。

【解释说明】

1. 黄连素在提取前应先把它切碎，研磨成粉状，否则会降低提取率。

2. 滤纸筒既要紧贴器壁，又要方便取放。被提取物高度不能超过虹吸管，否则被提取物不能被溶剂充分浸泡，影响提取效果。被提取物也不能漏出滤纸筒，以免堵塞虹吸管。

3. 减压蒸馏时，温度不可太高，否则产品会随乙醇一起蒸馏出去，降低产率。

4. 黄连素结晶应在50～60 ℃下干燥，如果温度过高，黄连素会变质或炭化。

5. 黄连素被硝酸等氧化剂氧化，转变为樱红色的氧化黄连素。黄连素在强碱中会部分转化为醛式黄连素，在此条件下，再加几滴丙酮，即可发生缩合反应，生成由丙酮与醛式黄连素缩合而成的黄色沉淀。

【操作注意事项】

减压蒸馏结束后，应先关闭热源，稍冷后缓慢解除真空，待系统内外压力平衡后再关闭真空泵，以防倒吸。

【废物处理】

黄连素回收利用，实验过程中产生的有机废液倒入专门的废液处理容器内。无机酸类废液不可以直接倒入下水道，应倒入专门的集中处理容器内。

【思考题】

1. 如果想对黄连素盐酸盐粗产品进行重结晶，应选用何种溶剂？

2. 在提取黄连素的过程中，为何要用石灰乳调节pH值？

4.3 菠菜叶中色素的提取和分离

菠菜叶中色素的提取和分离(操作视频)

【实验目的】

学习、掌握利用薄层色谱法、柱色谱法提取、分离天然产物的原理和基本操作。

【实验原理】

绿色植物(如菠菜叶)中含有叶绿素、胡萝卜素、水溶性维生素和叶黄素等多种天然产物。其中,叶绿素、β-胡萝卜素和少量的叶黄素是构成叶片颜色的主要组成。

叶绿素显绿色,存在两种结构相似的形式,即叶绿素 a(chlorophyll a,$C_{55}H_{72}N_4O_5Mg$)和叶绿素 b(chlorophyll b,$C_{55}H_{70}O_6N_4Mg$)。叶绿素 a 和叶绿素 b 都是卟啉(取代环四吡咯-卟吩衍生物)类化合物与金属镁的络合物,是植物光合作用所必需的催化剂。叶绿素 a 中一个甲基被甲酰基所取代从而形成叶绿素 b。叶绿素 a 和叶绿素 b 都易溶于乙醇、乙醚、丙酮、氯仿等有机溶剂。由于叶绿素中含有大的烃基结构,因此也易溶于醚、石油醚等非极性有机溶剂。叶绿素 a 是蓝黑色结晶,熔点为 150～153 ℃,其乙醇溶液呈蓝绿色,有深红色荧光。叶绿素 b 是深绿色粉末,熔点为 120～130 ℃,乙醇溶液呈绿色或黄绿色,有红色荧光,有旋光活性。叶绿素可用作食品、化妆品及医药的无毒着色剂。

胡萝卜素(carotenes,$C_{40}H_{56}$)是具有长链结构的共轭多烯(四萜),有三种异构体,即 α-胡萝卜素、β-胡萝卜素和 γ-胡萝卜素,其中 β-胡萝卜素含量最多,也最重要。生长期较长的绿色植物中,β-胡萝卜素的含量高达 90%。β-体具有维生素 A 的生理活性,在生物体内,β-体受酶催化氧化即可形成维生素 A。目前 β-体可工业生产,作为维生素 A 使用,也可用作食品色素。

叶黄素(lutein,$C_{40}H_{56}O_2$)是胡萝卜素的含氧衍生物(醇),在绿叶中的含量通常是胡萝卜素的两倍。与胡萝卜素相比,叶黄素更易溶于醇,但在石油醚中的溶解度较小。

图 4.3.1 为叶绿素 a、叶绿素 b、α-胡萝卜素、叶黄素和 β-胡萝卜素的结构式。

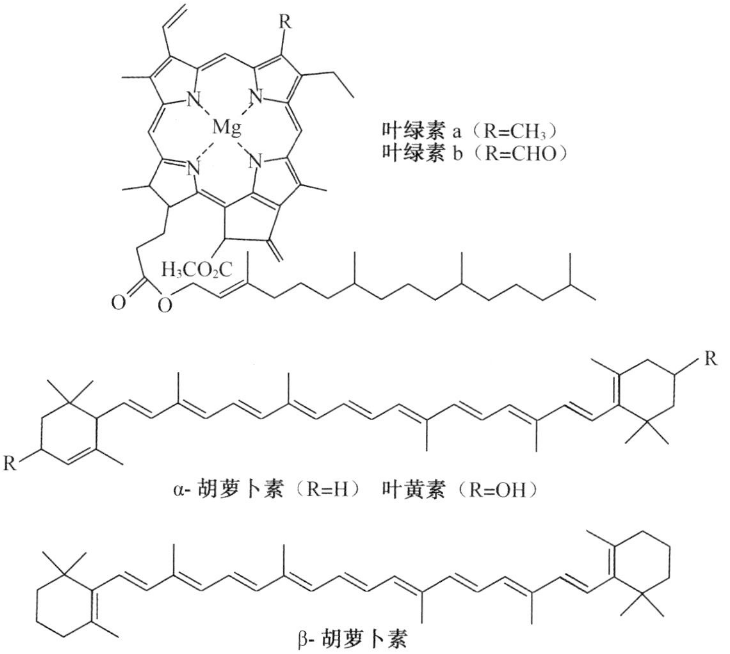

图 4.3.1 叶绿素 a、叶绿素 b、α-胡萝卜素、叶黄素和 β-胡萝卜素的结构式

(The structures of Chlorophyll a, Chlorophyll b, α-carotene, lutein and β-carotene)

薄层色谱(Thin-Layer Chromatography,TLC)分析是分离鉴定混合物、检测化合物纯度、跟踪反应的重要而有效的方法。柱色谱(Column Chromatography,CC)分析是分离、提纯有机物的重要方法。

本实验从菠菜叶中提取上述几种色素,并通过薄层色谱法和柱色谱法进行分离、鉴定和提纯。

【仪器与试剂】

仪器:层析柱(色谱柱),150 mL 锥形瓶,60 mL 梨形分液漏斗,小玻璃漏斗,玻璃棒,滴管,普通滤纸,脱脂棉,250 mL 烧杯,100 mL 量筒,25 mL 量筒,天平,铁架台。层析缸,玻璃板(10 cm×3 cm),毛细管。

试剂:正己烷,丙酮,无水乙醇,氯化钠,无水硫酸钠,硅胶(100 目),石英砂。

【实验步骤】

1. 色素提取

在研钵中放入 2 g 新鲜菠菜叶(剪碎的),研磨。加 22 mL 丙酮和 3 mL 正己烷,搅拌、研磨至叶片变白或者溶剂变为深绿色。用胶头滴管将研钵中的提取液转移至分液漏斗中,加入 20 mL 正己烷和 20 mL 10%氯化钠(wt%,即质量百分数)水溶液,振摇,静置。分去水层,有机层用 20 mL 10%氯化钠溶液水洗两次,然后将有机层用无水硫酸钠干燥。干燥好的提取液移至另一容器中。若溶液颜色较浅,可适当浓缩。

2. 薄层色谱分析

向干燥的层析缸加入展开剂[(正己烷 : 丙酮=7 : 3)(*V*/*V*)],盖好缸盖,摇动使其为溶剂蒸气所饱和。

取一块薄层硅胶板(10 cm×2.5 cm),在一端距边缘约 1 cm 处用铅笔划一横线,作为点样线(起始线),另一端距边缘适当距离(如 2 cm)处用铅笔轻划另一横线作为展开前沿(终点线)。

用毛细点样管吸取提取液,在点样线上轻轻点样(触点),若一次点样的斑点颜色较淡,待溶剂挥发后,重复点样,但斑点要尽量少。

待点样溶剂挥发后,将该薄层板以点样端向下置于层析缸中(浸入展开剂液面下约 0.5 cm)。盖好缸盖,静置。注意观察展开过程,当展开剂前沿上移至上端终点线时,立即取出。

待溶剂挥发,仔细观察并用铅笔圈画出每个斑点,量取并记下每个斑点展开的距离(量至斑点中心)。分别计算其比移值(R_f)。根据各斑点的颜色和比移值,尽可能多地鉴定出菠菜叶色素的各组分。

3. 柱色谱分析

装柱前应将色谱柱洗干净,进行干燥。在柱底层铺一小块脱脂棉,再铺约 0.5 cm 厚的石英砂,然后进行装柱。

将 20 g 硅胶与 50 mL 洗脱剂[正己烷∶乙醇＝20∶1(V/V)]在 100 mL 烧杯中混合，用玻璃棒搅拌成糊状。在色谱柱中装入 10 mL 洗脱剂，打开活塞，在色谱柱下面放一个干净并干燥的锥形瓶或烧瓶，缓慢接收洗脱剂。同时，将调好的硅胶缓慢装入色谱柱中。待所有硅胶全部装完后，用流下来的洗脱剂转移残留的硅胶，并将柱内壁残留的硅胶淋洗下来。柱子填充完后，在硅胶上端覆盖一层约 0.5 cm 厚的石英砂。在整个装柱过程中，柱内洗脱剂的高度始终不能低于硅胶最上端，否则柱内会出现裂痕和气泡。

加入样品前应先将柱内洗脱剂排至稍低于石英砂表面，然后移取 5 mL 菠菜叶提取液加入色谱柱中。样品加完后，打开色谱柱活塞，使液体样品进入石英砂层后，再加入少量洗脱剂将壁上的样品洗下来，待这部分液体进入石英砂层后，再加入洗脱剂进行淋洗。为了节省溶剂，本实验可先后采用 80 mL 正己烷-乙醇[20∶1(V/V)]、40 mL 正己烷-乙醇[10∶1(V/V)]、25 mL 正己烷-乙醇[5∶1(V/V)]进行梯度洗脱，直至所有色带被展开，分别收集不同色带，即得到菠菜叶中含有的不同色素。

【解释说明】

1. 薄层色谱分析中的终点线不划亦可。

2. 毛细点样管的管口要平整。

3. 点样时不可损坏硅胶层，以免影响展开。

4. 点样点不可浸入展开剂液面以下。

5. 若不划终点线，应在展开效果最好的时候取出，划下展开前沿。

6. 菠菜叶 TLC 分离一般可以显示四种颜色的 7 个斑点，分别是胡萝卜素(橙黄素)、脱镁叶绿素(灰色)、叶绿素 a 和叶绿素 b(蓝绿色和黄绿色，2 个点)以及叶黄素(黄色，3 个点)，也有观察到 8、9 甚至 10 个斑点的情况。菠菜叶 TLC 分离得到的各种色素比移值参考数据[展开剂：正己烷-丙酮 7∶3(V/V)，GF 硅胶板]如表 4.3.1 所示。

表 4.3.1 菠菜叶 TLC 分离得到的各种色素比移值参考数据

化合物	颜色	R_f
胡萝卜素	橙或黄色	0.93
脱镁叶绿素 a	灰色	0.55
脱镁叶绿素 b	灰色(或不可见)	0.47～0.54
叶绿素 a	蓝绿色	0.46
叶绿素 b	绿色	0.42
叶黄素及其他黄色素	黄色	0.41
	黄色	0.31
	黄色	0.17

【操作注意事项】

1. 层析薄层板应保护好硅胶层，用铅笔画线时不能破坏硅胶层。

2. 点样时毛细管要垂直，轻轻触及薄层板即可，点样量要适当。样品量太多，分离效果不好。

3. 柱色谱分析时，柱内洗脱剂的高度始终不能低于硅胶最上端。

【废物处理】

实验完成后，使用过的展开剂、洗脱剂应倒入专门的回收容器中进行回收处理。

【思考题】

1.点样薄层板展开时，点样点为什么不能浸入展开剂液面以下？

2.比较叶绿素、叶黄素和胡萝卜素三种色素的极性，为什么胡萝卜素移动最快、比移值最大？

【知识拓展】

视觉的化学——一个重要的异构化反应

4.4 醇、酚、醛、酮、羧酸未知液的分析

【实验目的】

1. 复习掌握醇、酚、醛、酮和羧酸的重要化学性质。

2. 学习和掌握用化学方法鉴定醇、酚、醛、酮和羧酸。

【实验原理】

官能团是决定有机化合物性质的原子或原子团，每种官能团都有其特征反应。通过分析官能团的化学性质，我们可以进行有机化合物的鉴定。有机官能团涉及的化学反应很多，这些化学反应若应用到有机化合物的分析鉴定中，应具备以下条件：①反应迅速；②有明显的化学现象，如颜色变化、溶解、沉淀、气体逸出等；③灵敏度高；④专一性强（指试剂与官能团反应专一）。

本实验要求根据实验室提供的试剂以及各化合物化学性质的不同，设计出一套用于鉴别指定化合物的流程。写出实验方案，经指导老师审阅后，方能进行试验。

【仪器与试剂】

仪器：烧杯，试管，试管架，试管夹，酒精灯，水浴锅。

试剂：5% $K_2Cr_2O_7$ 水溶液，浓 H_2SO_4，5% $KMnO_4$ 水溶液，1% $FeCl_3$ 水溶液，5%

$CuSO_4$ 水溶液，酚酞试液，10% NaOH 水溶液，多伦试剂，斐林试剂，5% 2,4-二硝基苯肼，I_2 溶液，饱和溴水，5% $AgNO_3$ 水溶液，5% $NaHCO_3$ 水溶液，浓氨水，蓝色石蕊试纸，水合茚三酮，Lucas 试剂。

【设计内容】

用化学方法分别鉴定下列各组化合物：

1. 正丁醇、仲丁醇、叔丁醇、乙二醇。

2. 苯酚、苄醇、苯丙氨酸、苯甲酸、水杨酸、苯甲醛。

3. 甲酸、乙酰乙酸乙酯、乙醛、丙酮、苯酚。

4. 葡萄糖、果糖、淀粉、蔗糖。

【实验步骤】

1. 设计实验方案：用给定的试剂预先设计出各组化合物的鉴定方案，写出实验步骤（可以画流程图）、预期现象，写出相关化学反应式。

2. 进行鉴别实验：实验过程中要认真观察和记录，对实验现象作出准确的分析判断，完成实验报告。

【解释说明】

1. 许多有机醇、酚和醚是有毒的，且具有易燃性，如丙酮是高度易燃的，只能在通风良好的环境使用这些化学品，并远离明火和其他火源。

2. 溴具有腐蚀性，会引起严重烧伤。使用时应小心，避免与皮肤、眼睛和衣服接触。如不慎接触，应用大量的水冲洗，严重者送医院诊治。

3. 托伦试剂久置后将生成 AgN_3 沉淀，容易爆炸，故需临时配用。实验时，切忌用明火直接加热托伦试剂，以免发生危险。实验完毕后，应加入少许稀硝酸，立即煮沸洗去银镜溶液。

4. 铬是剧毒的，其酸性溶液极具腐蚀性。为避免摄入，只能戴手套后处理。使用时小心避免接触皮肤、眼睛和衣服。若不慎接触，应用大量的水冲洗。若误食，立即就医。

5. 硝酸银有强氧化性，与皮肤接触，立即生成黑色的金属银，故滴加和摇晃时应小心操作，避免与皮肤接触。

【废物处理】

将所有废物弃置于相应标记的回收容器内，本实验中的任何有机化合物禁止倒入下水道！

【思考题】

1. 尽可能多地列出鉴别醛和酮的方法。

2. 有 5 瓶无标签的试剂，分别为甲酸、苯甲酸、苯甲醛、苯酚和苄醇，试选择合适的试剂鉴别它们。

3. 结合实验结果分析实验成败的关键，试说明本实验需改进的地方。

4.5 生物柴油的合成

【实验目的】

1. 学习生物柴油的合成原理及方法。

2. 了解生物质能和绿色能源。

【实验原理】

全球矿物能源储量十分有限，而全世界对于矿产能源的消费量却越来越大，开发可再生、环保的替代燃料已成为经济可持续发展最重要课题之一，生物燃料技术应运而生。生物柴油作为可替代石化柴油的清洁生物燃料，是一种生产成本和使用性能都与石化柴油基本相当，且具有良好的环境特性和可生物降解性的环保燃料，具有广阔的发展前景。

生物柴油的制备可采用物理法和化学法，物理法包括直接混合法和微乳液法，化学法包括高温热裂解法和酯交换法，其中酯交换法是目前合成生物柴油的主要方法。各种天然的植物油和动物脂肪以及食品工业废油都可作为酯交换生产生物柴油的原料。

在酯交换反应中，油料主要成分——甘油三酯与各种短链醇在催化剂作用下发生酯交换反应得到脂肪酸甲酯和甘油，可用于酯交换的醇包括甲醇、乙醇、丙醇、丁醇和戊醇，其中最常用的是甲醇，这是由于甲醇价格较低，碳链短，极性强，能够很快与脂肪酸甘油酯发生反应，且碱性催化剂易溶于甲醇。酯交换法包括酸催化、碱催化、生物酶催化和超临界酯交换法等，酯交换反应的反应式如下：

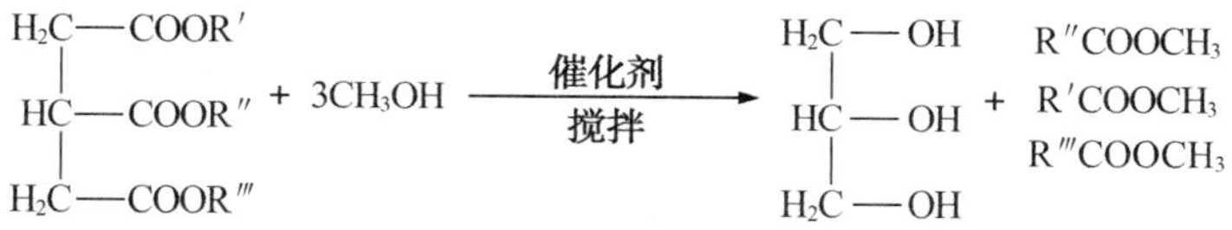

$$\begin{array}{l} H_2C-COOR' \\ \quad | \\ HC-COOR'' \\ \quad | \\ H_2C-COOR''' \end{array} + 3CH_3OH \xrightarrow[\text{搅拌}]{\text{催化剂}} \begin{array}{l} H_2C-OH \\ \quad | \\ HC-OH \\ \quad | \\ H_2C-OH \end{array} + \begin{array}{l} R''COOCH_3 \\ R'COOCH_3 \\ R'''COOCH_3 \end{array}$$

【仪器与试剂】

仪器：三口圆底烧瓶，搅拌器，球形冷凝管，温度计，水浴锅，铁架台，分液漏斗。

试剂：植物油，氢氧化钾，甲醇，盐酸溶液，无水硫酸钠，高碘酸，碘化钾，硫代硫酸钠，0.5%淀粉指示剂。

【实验步骤】

1. 生物柴油的合成

(1)图 4.5.1 为实验室制备生物柴油装置图，在装有冷凝管的 250 mL 三颈烧瓶中加入 100 g 大豆油，加热至 65 ℃后，边搅拌边加入 1.4 g 催化剂(KOH)和 21.8 g 甲醇溶液。恒温后开始计时，混合液充分反应 1 h 后，取出反应混合物，置于冰水浴中使反应及时结束。

(2)将反应混合物置于分液漏斗中静止分层，上层为黄色(即甲酯(生物柴油)和甲醇的混合物),下层为棕褐色(即甘油、未反应的甘油三酯)。收集上层液体,在 70 ℃下常压蒸馏,使甲醇与甲酯分离。

(3)蒸馏的残余物先用 36%盐酸洗涤一次,再用 85 ℃水洗涤数次,直至水相中无明显乳白色物质为止,除去脂肪酸盐、甘油以及水溶性物质、游离脂肪酸等。洗涤结束后,在残余物中加入足量无水硫酸钠,充分振荡,静置 10 min 后再过滤除去无水硫酸钠进行干燥,待油层变为浅黄色透明液体时,将生物柴油移出即可,称重并计算转酯率。

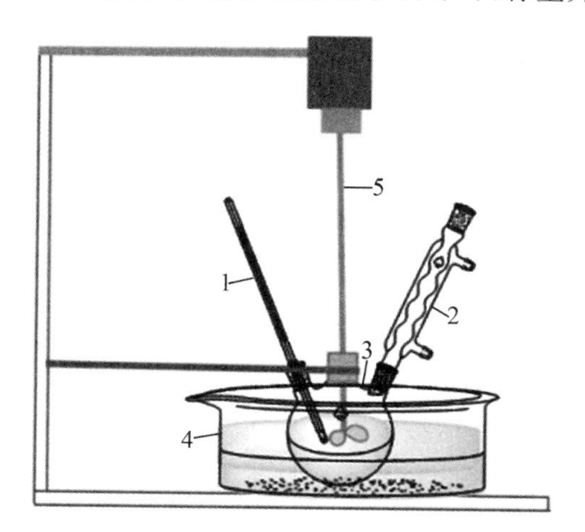

1—温度计(thermometer);2—球形冷凝管(Allihn condenser);3—三口圆底烧瓶(three-necked round bottom flask);4—水浴锅(water bath);5—搅拌器(stirrer)

图 4.5.1 实验室制备生物柴油装置图(A laboratory assembly for the preparation of biodiesel)

2. 合成产物中甘油含量的测定

(1)用电子分析天平称取 0.2 g 和 0.4 g(准确至 0.0001 g)生物柴油各一份,分别放入两烧杯中,加水溶解,冷却后分别转移至 100 mL 的容量瓶中,洗涤、定容待用。

(2)分别量取 25 mL 已配制的溶液于①号、②号 250 mL 的锥形瓶中,量取 25 mL 水于③号 250 mL 的锥形瓶中,再分别加入 20 mL 0.02 mol/L 的 KIO_4 溶液、10 mL 3 mol/L 的 H_2SO_4 溶液,盖好瓶盖,摇匀后在室温下黑暗处放置 30 min。

(3)然后加入 2 g KI、150 mL 水,析出的 I_2 用配置的 $Na_2S_2O_3$ 标液滴定,滴定至淡黄色时加入 1 mL 0.5%的淀粉指示剂,继续滴定,直到蓝色恰好消失为止,并做平行试验和空白试验。

【解释说明】

1. 酸催化法用到的催化剂为酸性催化剂,主要有硫酸、盐酸和磷酸等。在酸催化条件下,游离脂肪酸会发生酯化反应,且酯化反应速率要远快于酯交换速率,因此该法适用于游离脂肪酸和水分含量高的油脂制备生物柴油,其产率高,但反应温度和压力高,甲醇用量大,反应速度慢,反应设备需要不锈钢材料。工业上,酸催化法受到的关注程度远小于碱催化法。

2. 由植物油为原料制备生物柴油的产率为

转酯率=甲酯质量÷加入菜油的质量

3. 测定甘油含量的反应过程如下：

$$\text{氧化：}\begin{array}{l}H_2C—OH\\ \quad|\\ HC—OH\\ \quad|\\ H_2C—OH\end{array} + 2HIO_4 \longrightarrow 2HCHO + HCOOH + 2HIO_3 + H_2O$$

还原：$HIO_4 + 7I^- + 7H^+ = 4I_2 + 4H_2O$

$HIO_2 + 5I^- + 5H^+ = 3I_2 + 3H_2O$

滴定：$I_2 + 2Na_2S_2O_3 = 2NaI + Na_2S_4O_6$

【操作注意事项】

1. 在蒸馏过程中,温度计的水银球应与支管下口处于同一水平位置,升温过程不易过快。

2. 用无水硫酸钠干燥产品时,完全干燥应做到干燥剂既不沾壁又不黏结成块,同时干燥剂不宜加多,否则会导致产品被吸附,造成损失。

3. 配制标准溶液时,应注意正确使用容量瓶;定容时,视线应与刻度线齐平。

4. 滴定时,应使滴定管尖嘴部分插入锥形瓶瓶口(或烧杯口)下 1～2 cm 处,滴定速度不能太快,以每秒 3～4 滴为宜,切不可成液柱流下。边滴边摇晃,临近终点时,应一滴或半滴地加入,慢滴快搅,30 s 不变色即为滴定终点。

【废物处理】

将剩余的甘油三酯和制得的生物柴油分别回收,实验过程中产生的废液倒入专门的废液处理容器内,不得直接倒入下水道。无机酸类废液也应回收到相应的处理容器内,不得直接倒入下水道。

【思考题】

1. 根据实验结果,如何计算产物中甘油的含量?

2. 为什么碱催化酯交换法中通常选用 KOH 作为碱催化剂?

4.6 甲基橙的制备

【实验目的】

1. 掌握重氮化反应和偶合反应的实验操作。

2. 巩固盐析和重结晶的实验操作。

【实验原理】

甲基橙(对二甲基氨基偶氮苯磺酸钠)又名金莲橙 D,广泛应用于生产和科学实验,常用作酸碱指示剂和生物染料，还可用于印染纺织品和分光光度测定氯、溴、和溴离子。

甲基橙是由对氨基苯磺酸重氮盐与N,N-二甲基苯胺的醋酸盐在弱酸性介质中偶合得到的，偶合首先得到的是嫩红色的酸式甲基橙（称为“酸性黄”），在碱中酸性黄转变为橙色的钠盐，即甲基橙，如图4.6.1所示。甲基橙稍溶于水（呈现黄色），易溶于热水，几乎不溶于乙醇。

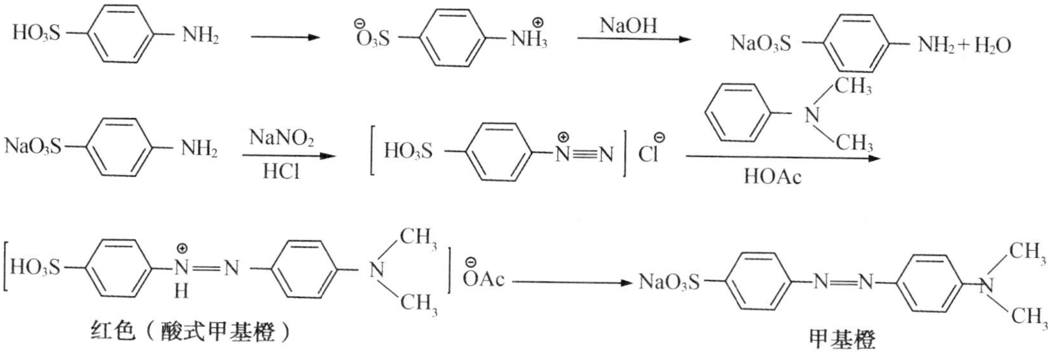

$HO_3S-C_6H_4-NH_2 \longrightarrow {}^{\ominus}O_3S-C_6H_4-NH_3^{\oplus} \xrightarrow{NaOH} NaO_3S-C_6H_4-NH_2 + H_2O$

$NaO_3S-C_6H_4-NH_2 \xrightarrow[HCl]{NaNO_2} [HO_3S-C_6H_4-\overset{\oplus}{N}{\equiv}N]Cl^{\ominus} \xrightarrow[HOAc]{C_6H_5N(CH_3)_2}$

$[HO_3S-C_6H_4-\overset{\oplus}{N}H{=}N-C_6H_4-N(CH_3)_2]{}^{\ominus}OAc \longrightarrow NaO_3S-C_6H_4-N{=}N-C_6H_4-N(CH_3)_2$

红色（酸式甲基橙）　　　　甲基橙

图4.6.1 甲基橙的制备反应式（The main reaction of the preparation of methyl orange）

【仪器与试剂】

仪器：烧杯，试管，布氏漏斗，锥形瓶，抽滤瓶，玻璃棒。

试剂：对氨基苯磺酸晶体，5%氢氧化钠溶液，亚硝酸钠，浓盐酸，N,N-二甲苯胺，冰醋酸，乙醇，乙醚，淀粉-碘化钾试纸，pH试纸，冰水。

【实验步骤】

1. 重氮盐的制备

(1)在100 mL烧杯中放置对氨基苯磺酸晶体2.0 g(0.0115 mol)，加入5%氢氧化钠溶液10 mL，温热溶解后冷却至室温。将0.8 g(0.0116 mol)亚硝酸钠溶于6 mL水，加入上述溶液，冰水浴冷却至0～5 ℃。

(2)再将3 mL浓盐酸用10 mL水配成的稀盐酸溶液，缓慢滴入到冷却的混合溶液中，边滴边搅拌，保持温度在5 ℃以下，滴加完后，用淀粉-碘化钾试纸检验，试纸应为蓝色（若不显蓝色，再补加亚硝酸钠溶液），继续在冰水浴中搅拌15 min，可观察到有白色细粒状的重氮盐析出。

2. 偶合反应

(1)将1.3 mL(1.248 g,0.0103 mol)N,N-二甲苯胺和1 mL冰醋酸在试管中混匀，再将它们缓慢加入制得的重氮盐的冷悬浊液中，滴完后继续在冰水浴中搅拌10 min，使其偶合完全。在这过程中，可观察到有红色沉淀析出。

(2)在搅拌下，缓慢向混合液中滴加5%氢氧化钠溶液（约25 mL），直至pH试纸显碱性，沉淀由红色转为橙色。再将反应混合物在沸水浴上加热5 min，冷却至室温以下，再用冰水冷却，使甲基橙晶体析出完全。抽滤收集晶体，用少量水洗涤，再依次用少量乙醇、乙醚洗涤，压干，即得甲基橙粗产品。

3. 纯化

将甲基橙粗产品溶于稀氢氧化钠溶液中，加热溶解，冷却结晶，抽滤，再依次用少量乙醇、乙醚洗涤，干燥即得甲基橙成品。

4. 甲基橙的鉴定

取少量甲基橙成品溶于水中，溶液呈现橙色，加入稀盐酸后，溶液变为紫红色，再加入氢氧化钠稀溶液后，溶液变为橙色。

【解释说明】

1. 重氮化过程中，应严格控制温度，反应温度若高于 5 ℃，生成的重氮盐易水解为酚，降低产率。

2. 淀粉-碘化钾试纸刚变蓝说明亚硝酸刚过量，重氮化完全，否则再加亚硝酸钠水溶液。用淀粉-碘化钾试纸检验，若亚硝酸过量，会引发副反应，应加尿素除去。

3. 当滴加氢氧化钠溶液至溶液变为橙色时应立即停止加入，因为未反应完全的 N，N-二甲基苯胺醋酸盐在碱性下可析出难溶于水的 N，N-二甲基苯胺，影响产品纯度。

4. 由于产物呈碱性，温度高易变质，颜色变深，所以重结晶时应迅速，反应物在水浴中加热时间不能太长，温度不能太高。用乙醇、乙醚洗涤的目的是使晶体迅速干燥，湿的甲基橙在空气中受光照后颜色会很快变深。

【操作注意事项】

1. 本反应温度控制十分重要，制备重氮盐时，温度应保持在 5 ℃以下。

2. 反应产物在水浴中加热时间不能太长（约 5 min），温度不能太高（60～80 ℃），否则产物变质，产品颜色变深。

3. 由于产品晶体较细，抽滤时应防止将滤纸抽破（布氏漏斗不必塞得太紧）。湿的甲基橙接受日光照射也会变质，通常用乙醇、乙醚洗涤，55～78 ℃烘干。所得产品为一种钠盐，无固定熔点，不必测定。

【废物处理】

回收产品甲基橙晶体，有机废液和无机废液分别集中倒入相应的废液处理容器内，不得直接倒入下水道中，用完的浓盐酸也应及时封好。

【思考题】

1. 重氮盐为什么可以与酚或胺偶合呢？偶合条件又分别是什么？

2. 本实验中，制备重氮盐时，为什么要把对氨基苯磺酸变成钠盐？本实验若改成先将对氨基苯磺酸与盐酸混合，再加亚硝酸钠溶液进行重氮化反应，可以吗？

3. 试解释甲基橙在酸性介质中变色的原因，并用反应式表示。

4. 何为重氮化反应？重氮化反应的条件是什么？

4.7　基于分子模拟的有机化学实验

基于分子模拟的
有机化学实验

基于分子模型的有机化学
实验(实验讲解)

第 5 章　文献实验

文献实验

参考文献

[1] 庞华，郭今心. 有机化学实验[M]. 济南：山东大学出版社，2006.

[2] 兰州大学编. 有机化学实验[M]. 王清廉，李瀛，高坤，等修订. 4 版. 北京：高等教育出版社，2017.

[3] 李吉海，刘金庭. 基础化学实验Ⅱ[M]. 北京：化学工业出版社，2007.

[4] 贾瑛，许国根，张剑. 绿色有机化学实验[M]. 西安：西北工业大学出版社，2009.

[5] 唐玉海，刘芸. 有机化学实验[M]. 2 版. 北京：高等教育出版社，2020.

[6] 薛思佳，季萍等. 有机化学实验(英汉双语)[M]. 3 版. 北京：科学出版社，2016.

[7] 柏一慧. 有机化学实验(英汉双语)[M]. 北京：科学出版社，2019.

[8] 龙盛京. 有机化学实验[M]. 2 版. 北京：人民卫生出版社，2011.

[9] 丁长江. 有机化学实验[M]. 2 版. 北京：科学出版社，2016.

[10] 冯文芳. 有机化学实验(英汉双语)[M]. 武汉：华中科技大学出版社，2013.

[11] MOHRIG J R, ALBERG D, Hofmeister G, et al. Laboratory Techniques in Organic Chemistry [M]. 4th ed.New York: W.H. Freeman and Company, 2014.

[12] PAVIA D L, LAMPMAN G M, KRIZ G S, et al. Introduction to Organic Laboratory Techniques: A Microscale Approach [M].4th ed. Belmont: Thomson Higher Education, 2007.

[13] WILLIAMSON K L, MASTERS K M. Macroscale and Microscale Organic Experiments[M]. Boston: Cengage Learning, 2017.

[14] MUKHERJEE A. Integration of Fundamental Organic Chemistry with Green Chemistry, A Laboratory manual[M]. Oxford: Alpha Science International LTD, 2019.

[15] TANG K, LI Q. Biochemistry of Wine and Beer Fermentation [M]. // ASHOK P, MARIA Á S, GUOCHENG D, et al. Current Developments in Biotechnology and Bioengineering. Amsterdam: Elsevier, 2017, 281-304.

[16] GIBSON M, NEWWSHAM P. Taste, Flavor and Aroma[M]. //MARK G,

PAT N. Food Science and the Culinary Arts. Amsterdam:Elsevier,2018:35-52.

[17] 杨华,刘怡虹,梁国友,等. 从黄连中提取黄连素[J]. 广州化工,2012,40(12):79-80.

[18] 张来新,杨琼,李小卫. 黄连中提取黄连素[J]. 贵州化工,2003,28(2):30-32.

[19] 龙城宇. 甲基橙的制备[J]. 工业,2016,6 (23):274.

[20] ZHU G,YU G. A pineapple flavor imitation by the note method[J]. Food Science and Technology,2020,40: 924-928.

附录

附录Ⅰ　常用元素的质量

常用元素的质量如附表1.1所示。

附表1.1　常用元素的质量

元素名称	元素符号	原子量/(g/mol)	元素名称	元素符号	原子量/(g/mol)
银	Ag	107.868	碘	I	126.905
铝	Al	26.982	钾	K	39.098
溴	Br	79.904	镁	Mg	24.305
碳	C	12.011	锰	Mn	54.938
钙	Ca	40.078	氮	N	14.007
氯	Cl	35.453	钠	Na	22.990
铬	Cr	51.996	氧	O	15.999
铜	Cu	63.546	磷	P	30.974
氟	F	18.998	铅	Pb	207.200
铁	Fe	55.847	硫	S	32.066
氢	H	1.008	锡	Sn	118.710
汞	Hg	200.59	锌	Zn	65.390

附录Ⅱ　常用酸、碱溶液的相对密度及质量分数

常用酸、碱溶液的相对密度及质量分数

附录Ⅲ　常用试剂的配制

常用试剂的配制

附录Ⅳ　常见有机化合物的理化性质

常见有机化合物的理化性质

Chapter 1 Fundamentals of Organic Experiments

1.1 General Rules for the Organic Chemistry Laboratory

Experimental safty education video

Organic chemistry is an experimental science. Organic chemical knowledge has accumulated from experimental observations and studies made by thousands of scientists. Organic chemistry experiment course is intended to introduce us to major concepts and techniques in organic chemistry through laboratory experiments. In the laboratory, we examine, test, and establish for ourselves the chemical principles studied in class and from text books, collect experimental data, and use our reasoning to draw logical conclusions about the meaning of these data. Therefore, it is not only in the laboratory that we learn "how we know that we know", we also learn the expanding knowledge base and critical thinking skills which will help us to prepare for a wide array of potential future challenges, including the upper level courses, organic requirements for medical schools, and independent research.

The following guidelines help students to perform chemistry experiments safely in laboratories. Students should follow the safety instructions given below in chemistry laboratory.

1. Everyone should carefully preview the experimental content, understand the experimental method and experimental principles, and complete the preview report prior to arriving at the laboratory. In experimental operation, operating procedures must be strictly followed. Understand the hazards and safe operation methods of reagents, and learn the operating procedures of relevant experimental instruments and equipment.

2. Dress sensibly in the laboratory. Wear laboratory coats which are made of cotton. Wear shoes, and don't wear sandals, slippers or high heels. Confine long hair and loose clothes. Don't wear shorts, tank tops or skirts.

3. Be familiar with laboratory environment and safety rules. Know where to find and how to use water, electricity, safety showers, eyewashes, fire extinguishers, fire blanket and first aid kit. Report injuries, accidents and other incidents to your instructor and follow his or her instructions for treatment. Necessary safety measures should be taken according to the experimental situation, such as wearing goggles, masks and gloves.

4. Before the start of an experiment, check all glass equipment. During the experiment, use it carefully and skillfully. If the equipment is damaged, report it to the teacher in time.

5. Before using the chemicals, read their labels carefully. Use them only as required dosage in the experiment with caution and don't waste them. Never pour any reagent back into a stock bottle. Chemical bottles should be put back to the original place and covered with the stopper immediately after use and avoid the stopper being confused as well as the chemicals being contaminated. Don't change the position at random of normal reagents and common instruments in the lab such as balance, fire extinguisher, distilled water bucket and so on.

6. Students may not redo the experiment or change the experimental scheme. Report all abnormal conditions to your teacher to minimize the operational hazards. Everyone should follow the teacher's instructions, work carefully, observe carefully, think positively and record truthfully.

7. Your full attention must be given to what you are doing during the experimental period. Never leave an ongoing experiment unattended. No cell phone usage is allowed in the lab. Never eat, drink or smoke in the laboratory. Don't do anything unrelated to the experiment. Don't leave the lab without permission.

8. During the experiment, keep your experimental area and whole lab tidy. All kinds of solid or liquid waste should be placed in authorized containers. Match stems, waste filter papers and other waste must not be thrown into the sink to prevent blockage of sewer.

9. Properly dispose of chemical waste. All kinds of waste should be classified and recycled in different waste disposal containers like organic waste, aqueous waste, solid waste and broken glassware.

10. Students on duty should stay to clean up the laboratory. Before leaving the laboratory, check carefully whether water and power are switched off safely. Wash your hands.

Overall, you are responsible for developing good laboratory habits.

1.2 Prevention and Treatment of Accidents in the Laboratory

Chemicals must always be handled with great care and attention. Many are potentially dangerous to health, possibly even poisonous, flammable, corrosive and irritating and some are explosive. And chemical reactions are often carried out in different circumstances with various heating source, electrical appliances, glassware or other equipment. Occasional accidents will cause fire, explosion, electric shock, cuts, burns or poisoning. But these accidents can be prevented and avoided as long as we follow the rules strictly when we do experiments. In order to prevent and deal with dangerous accidents, you should be familiar with the fundamentals of laboratory safety.

Always inform your instructor immediately of any safety incident or accident that happens to you or your neighbors.

Remember that serious accidents are irreversible and everyone only lives once.

1.2.1 Fire prevention and treatment

1.2.1.1 Prevention

1. Flammable substances are the most common hazard of the organic laboratory, therefore, particular care should be taken when handling flammable solvents such as benzene, acetone, petroleum ether, carbon disulfide or alcohol, etc. which should be kept away from an open flame. Especially when diethyl ether is used, indoor flame is strictly prohibited. Overall, pay close attention to sources of ignition, such as open flame, sparks, and hot surfaces.

2. Never allow any vapor of a volatile chemical to escape into the open laboratory. If spillage of solvent or accidental release of flammable vapor occurs, the whole laboratory should be ventilated as soon as possible. Work in an exhaust hood when manipulating large quantities of flammable liquids.

3. Don't heat flammable liquids over an open flame when refluxing or distillation. Use a water bath or oil bath to heat them while keeping the cooling water flow smooth

through the condenser from the lower end to the upper end during the experiment.

4. Flammable liquids should not be heated in beakers or open containers over an open flame.

5. For refluxing or distillation, one or two boiling stones should be added in the distillation bottle to prevent superheating of the liquid and reduce the tendency of the liquid to "bump". If you don't add any boiling stones when starting distillation, stop heating immediately and re-add them after cooling.

6. Flammable, explosive and volatile chemicals mustn't be discarded randomly or poured into the waste tank, instead, they should be poured into appropriately labeled recovery bottles. These bottles should be placed in the hood in the lab.

1.2.1.2 Treatment

In case of a fire, you should be calm. First of all, cut power and the gas off, move the flammable and explosive reagents away, and then take correct measures to control the fire. If the fire is out of control, call 119 fire alarm in time.

1. If it is a small fire in a flask, or a beaker, it usually can be extinguished quickly by placing a flat object such as an asbestos pad, a glass plate or a watch glass over the mouth of the flask or beaker to cut off the supply of air.

2. If the burning liquid is spilled on the floor or on the table, apply dry fine sand to extinguish it. Water mustn't be used to extinguish burning liquids such as benzene, petroleum ether, etc. that is lighter than water, because the burning liquid will spread on the water surface, and enlarge the combustion area.

3. When a person's clothing catches fire, shove the person down and roll him or her body tightly in a fire blanket over to extinguish the flames. It is extremely important to prevent the victim from running or standing because the greatest harm comes from breathing the hot vapors that rise past the mouth. Dumping a huge volume of water in a short period of time by the nearby safety shower is also effective.

4. In case of a large fire, use a fire extinguisher in the premise of ensuring own safety. Carbon dioxide extinguisher is a common fire extinguisher in organic chemistry labs, which is predominantly used to put out the fire of electric appliance or grease and other organic compounds. When using a carbon dioxide extinguisher, the carbon dioxide gas is ejected from the horn of the extinguisher. Do not hold onto the horn because it will become extremely cold. Note that different types of fire require different types of fire extinguishers. Do not use water to extinguish chemical fires. To use a fire extinguisher, aim low and direct the nozzle first toward the edge of the fire and then

toward the middle.

5. Never use water or carbon tetrachloride extinguishers to extinguish burning sodium and potassium. Because sodium and potassium can react violently with both water and carbon tetrachloride, usually dry fine sand is used to cover them to put fire out.

The performance and applicable scope of commonly used fire extinguishers are shown in Table 1.2.1.

Table 1.2.1 Commonly used fire extinguishers

Type of fire extinguisher	Composition	Scope of application & characteristics
Carbon dioxide extinguishers	Liquid CO_2	Electrical appliance, combustible and flammable liquid fire
Water-based foam extinguishers	Foam AFFF and nitrogen gas	Flammable liquid fire such as petrol, paint, oil, etc.
Dry power extinguishers	The main components are sodium bicarbonate and other salts mixed with a proper amount of lubricant and moisture proof agent	Fire within seconds of oil, flammable gas, electrical equipment, precision instruments, books and documents, etc.
Clean gas fire extinguishers	heptafluoropropane, trifluoromethane, IG541 (argon, nitrogen, carbon dioxide)	Fire of oil, organic solvents, precision instruments and high-voltage electrical equipment

1.2.2 Explosion prevention

1. First and foremost, always wear protective glasses or goggles. If students wear contact lenses, they should inform teachers and wear the safety glasses throughout the lab session.

2. Never heat a closed system or conduct a reaction in a closed system.

3. The apparatus should be assembled correctly. The whole system should not be closed in the process of normal distillation and reflux. Do not use flat-bottomed or thin-walled glassware in a vacuum distillation. Always check flask for defects such as star cracks. Distillation to dryness is dangerous.

4. Do not knock or press the explosive solids heavily such as acetylene metal salts, picric acid, picric acid metal salts, trinitrotoluene (TNT) and so on to avoid the explosion. The small amount of the residue is not allowed to throw away. It should be put into appropriately labeled recovery bottles.

5. A fierce explosion or combustion can be produced when some organic compounds come into contact with strong oxidizing agents such as nitric acid, permanganate ion, and peroxides; alkali metals such as sodium; or very finely divided metals such as zinc dust and platinum catalysts. Beware of their handling and storage.

6. Keep flammable organic solvents (especially low boiling flammable solvents) away from open flames. Do not pour flammable solvents into waste tanks, and do not store flammable solvents in open bottles.

7. The presence of peroxides must be checked and disposed with ferrous sulfate before using the ether.

8. If the reaction is too violent, it is better to control the feeding speed and reaction temperature. And cooling measures should be taken if necessary.

1.2.3 Electric shock prevention

Electrical equipment should be checked carefully whether there is open circuit or leakage prior to power connection, in case of minor electrical shock, the power supply should be immediately cut off, and the equipment must be inspected and repaired. When using electrical appliances, hand, clothes and devices must be dry. Avoid direct contact between human body and the conductive part of the instrument. All these devices should be connected through ground-fault interrupters to minimize the shock hazards. The power should be cut off or unplugged after the experiment.

1.2.4 Cut prevention and treatment

1.2.4.1 Prevention

Many cuts occur when handling glass. Many accidents can be prevented if glass is handled with a glass cloth or leather gloves. To avoid cuts, first never place glassware under pressure that it is not designed to withstand and pay attention to the following points.

1. When inserting glass tubing into a cork or rubber stopper, slowly rotate the glass tubing to prevent cutting, sometimes lubricate with a drop of glycerin and protect hands with a towel.

2. Do not use glassware that has cracks or cracked. Wash glassware with great care to avoid breaking.

3. Glassware such as reagent bottles, cylinders and watch glass must not be heated. Hot glassware cannot be placed on a cold surface or added cold water suddenly to avoid rupture.

4. Do not put broken glass in the trash can. Dispose of them in the broken glass container.

1.2.4.2 Treatment

Learn the location of the first aid kit in your lab for the treatment of simple cuts. All injuries, no matter how slight, should be reported to the teacher immediately.

Small cuts should be rinsed thoroughly with running water. Glass chips in the wound should be all removed with sterilized tweezers. Then the wound should be wiped some iodine ethanol solution and tied the bandage. If the cut is severe, bind up with gauze to slow down the bleeding, and then go to the hospital for further treatment.

1.2.5 Burns prevention and treatment

1.2.5.1 Prevention

Skin may be burned after contacting with hot, cold or corrosive substances. All students are required to wear a lab coat at all times. Wear safety goggles and gloves when handling hazardous chemicals to avoid burns.

1. Special care should be taken when handling hot objects and corrosive chemicals to avoid their contact with your body.

2. When the liquid in a test tube is heated or boiled, the nozzle should not be directed at oneself or others; and one is not allowed to approach to the mouth of the test tube or the flask to observe the reaction in it.

3. Don't pour water into the acid. To dilute concentrated acid H_2SO_4, add acid carefully to water (not vice versa), and stirring at the same time. Wear goggles when heating concentrated acid or alkali solution.

1.2.5.2 Treatment

For a heat burn, apply cold water for 10-15 min. Seek immediate medical attention for any extensive burn. For a cold burn, do not apply heat. Instead, treat the affected area with large volumes of warm water and seek medical attention.

1. If the burn is more serious such as black skin, larger burns etc., it should be covered with disinfect gauze or with cloths soaked in cold water. These procedures may

reduce the pain. Then immediately seek professional medical attention. Do not cover burns with oily or greasy ointments.

2. Minor burns can be covered with picric acid ointment or vanillin with some tannic acid.

3. If the skin is burned by acids, first rinse the affected area immediately with large quantities of water, then wash with 1%-3% sodium bicarbonate solution, and then rinse with water.

4. If the skin is burned by alkali, first rinse the affected area immediately with water, then wash with 1% aqueous acetic acid or 3% aqueous boric acid and then rinse with water.

5. If bromine or phenol drops on the skin, use alcohol to wash the affected area, and then coat it with glycerol.

6. If the acid gets into the eye, it should be washed immediately with water, then washed with 1% sodium bicarbonate solution, and then rinse with water.

7. If the alkali gets into the eye, it should be washed immediately with larger quantities of water to ensure thorough washing, then washed with 1% aqueous boric acid or 0.5% acetic acid solution and then rinse with water.

1.2.6 Poison prevention and treatment

Knowledge of the toxicity of chemical substances is an essential part of the training of a responsible chemist. Many chemicals have special toxicity. Poisoning can be caused by the inhalation of toxic gases or eating toxic substances. Some toxic chemicals can also penetrate into the body from cuts or burns. In order to prevent poisoning, we should pay attention to the following points.

1. Toxic drugs should be kept properly.

2. Avoid skin contact by wearing the proper type of protective gloves and avoid inhalation by working in a good exhaust hood.

3. Wash hands immediately after the experiment.

If you have some poisoning symptoms such as dizziness, headache, or other symptoms during the experiment you should leave the laboratory area and move to an area where you can breathe fresh air and then seek medical treatment.

Medical treatment is easier if the nature of the poison can be established (obtain a sample of the poison, gas, vomit, etc.).

1.3 Disposal of Waste in the Laboratory

Organic experiments always generate different kinds of wastes. Waste disposal has been one of the major environmental problems nowadays. The useful handling of lab wastes can be done in the following ways.

1. All laboratory waste cannot be discarded at random. It can be classified into solid or liquid waste, hazardous or nonhazardous waste, and disposed of properly. Some hazardous wastes that are difficult to handle can be sent to the environmental protection department for special treatment.

2. Broken glassware and flammable materials (such as waste paper or fabric that has been used to wipe flammable liquids) should be placed separately in different trash cans with lids. Gloves, masks, syringes, etc. should be disposed of as medical waste in special waste bins.

3. Nonhazardous solid waste (such as corks, silica gel and magnesium sulfate) is placed in the ordinary trash can. Toxic solid waste should be sealed with plastic bags and placed in a clearly labeled container.

4. Waste organic solvents should be poured into properly labeled waster-solvent containers and stored in a well-ventilated place.

5. In particular, halogenated solvents (such as dichloromethane) or solvents containing halides should have special containers for recovery. The waste is transported to the specialized processing agencies.

6. Waste aqueous solutions are collected separately from organic waste to prevent violent reactions between them, and they are stored and handled differently.

7. Some carcinogenic substances must be handled with great care, avoiding contact with your body.

In any case, it is not permitted to dispose of solvents by pouring them down the sink.

Further Reading

Twelve principles of green chemistry

1.4 Mostly Used Glassware and Apparatus in Organic Laboratory

1.4.1 Mostly used glassware

The common types of glassware can be divided into non-standard taper glassware (Figure 1.4.1) and standard-taper ground glassware (Figure 1.4.2).

1.4.1.1 Common glassware

A typical set of lab glassware for organic experiments is shown in Figure 1.4.1.

1.4.1.2 The standard-taper ground glassware

The standard taper is equipped with a ground glass fitting that is usually labeled in different sizes with different numbers such as 10#, 14#, 19#, 24#, 29#, 34#, and 40#, etc. These numbers represent the outer diameter (OD) in millimeters (mm) at the widest point of the inner joint. The ground glass joints can also be labeled with a number, a slash, and another number. The first number represents the OD at the widest point of the inner joint. The second number represents the ground glass length of the joint in millimeters. For example, a 19/22 joint is 19 mm wide at the top (the widest part) and is 22 mm in length. Any male joint will fit a female joint. [When it is necessary to connect two glassware with different joint sizes, a connecting adapter (reducing adapter or enlarging adapter) is available to place between them to make the connection.] Condensers, distillation heads, claisen adapters, vacuum adapters, separatory funnels, and round-bottomed flasks are commonly fitted with these joints as shown in Figure 1.4.2.

1.4.2 Mostly used apparatus

Mostly used apparatus for organic chemistry experiment includes reflux, distillation and fractional distillation apparatus, etc. as shown below.

1.4.2.1 Reflux apparatus

The reflux apparatus allows an organic reaction to be carried out at the boiling point of the solvent and yet prevents loss of solvent or reagent due to evaporation. Common reflux apparatus is shown in Figure 1. 4. 3. If the reaction needs to be moistened, a drying tube can be connected to the upper end of the spherical condensation tube, as shown in Figure 1.4.3 (b). If water-soluble gases such as HCl and

SO_2 are generated at reflux, the gas absorption device can be connected, as shown in Figure 1.4.3(c). In Figure 1.4.3(d),(f),(g), liquid can be added at the same time during reflux. Figure 1.4.3(f) shows a Soxhlet extractor. The sample to be tested is wrapped in filter paper and placed in the extraction tube, and the lipid substances in the sample are extracted by reflux.

1.4.2.2 Distillation apparatus

Simple distillation apparatus and vacuum distillation apparatus are shown in Figure 1.4.4. Distillation is often used to separate two or more liquids with a large difference of more than 25 ℃ in boiling point, and can also be used to remove organic solvents. A general distillation apparatus is shown in Figure 1.4.4 (a), which usually uses water-cooled condensing pipe, and the condensed water flows from the bottom to the top. If the boiling point of the distilled liquid is above 140 ℃, the straight water-cooled condensing pipe can be replaced with air condensing pipe to avoid cracking the condensing pipe due to excessive steam temperature. A vacuum distillation apparatus is shown in Figure 1.4.4 (b), which is commonly used for distillation of substances easy to decompose, oxidize, polymerize, etc. If the solvent needs to be added in the distillation process, a drip funnel can be added at the upper end of the distillation head.

1.4.2.3 Fractional distillation apparatus

Fractional distillation apparatus is shown in Figure 1.4.5. When the boiling points of the two liquid components in the mixture are small and cannot be separated by distillation, accurate separation is required by fractional flow method. Spherical fractionating column, Wechsler fractionating column (spiked fractionating column) and packed fractionating column are shown in Figure 1.4.5(a), (b) and (c). The common fractional distillation apparatus is shown in Figure 1.4.5(d).

1.4.2.4 Gas absorption apparatus

Common gas absorption apparatus is shown in Figure 1.4.6. They are used to absorb soluble gases produced in the reaction, such as HCl, SO_2, etc. The device in Figure 1.4.6 (a),(b) can be used to absorb small amounts of gas. In Figure 1.4.6 (a), the funnel is tilted, with half of it submerged in water to prevent gas spillover,and half of it above the water to prevent water from being sucked back into the reaction bottle. Figure 1.4.6 (c) can be used when a large amount of gas is produced or when reaction water is produced at a fast rate flows into the suction flask from the upper end and spills out at a constant liquid level. The glass nozzle is below the water's surface, preventing gas from escaping into the atmosphere.

1.4.2.5 Stirring device apparatus

Common stirring apparatus is shown in Figure 1.4.7. The stirring apparatus can make the heterogeneous reaction mix quickly and evenly to prevent uneven concentration or local overheating phenomenon, avoiding the occurrence of side reaction. The apparatus in Figure 1. 4. 7 (a) is a reflow device with stirring and temperature measurement. The apparatus can drop liquid in the stirring process in Figure 1.4.7 (b). The device can stir, measure temperature, reflux and add liquid at the same time in Figure 1.4.7 (c).

1.4.3 Precautions for using glassware

1. All glassware should be used carefully to avoid impact or breakage.

2. Don't heat glassware directly except boiling flasks and tubes.

3. Erlenmeyer flask and flat-bottom flask cannot withstand reduced pressure and should not be used in such systems.

4. After cleaning up glassware containing a stopper, a small piece of paper must be put between the stopper and ground joint to avoid adhesion in next use.

5. If a joint becomes frozen, it can sometimes be loosened by tapping it gently with the wooden stick. If this procedure fails, you may try heating the joint in hot water or a steam bath. Hair dryer can also be used to heat the joint.

6. The piston and the stopper of the separatory funnel are both ground-glass, if not original, they may not be tight. So pay much attention to the protection of them.

7. The glass of a mercury bulb is thin and easy-to-break thus should be used with care. Never use it as a stirring rod! After use, cool it down, and rinse it afterwards to keep away from cracking. The measurement of thermometer doesn't go beyond its graduated range.

8. If the mercury thermometer is accidentally broken, immediately cover the spilled mercury with sulfur powder, and insert the broken thermometer into a container with a lid containing sulfur.

1.5 Cleaning and Drying Glassware

1.5.1 Cleaning glassware

As the degree of cleanliness of the glassware used in the experiment directly affects the experimental results. Therefore, students should learn how to clean and dry the glassware, and develop the habit of cleaning the glassware immediately after each experiment.

Generally, the glassware can be washed with water firstly, and then washed with scouring powders or detergent using brush, and the final rinse can be done with distilled or deionized water. The glassware has been cleaned when the glassware is upside down and there are no water droplets on the wall.

Never use chemical reagent or organic solvent thoughtlessly to rinse glassware. This may produce waste, and create a hazardous situation, resulting in pollution to our environment.

It is also convenient and quick to clean glassware by the ultrasonic oscillator.

1.5.2 Drying glassware

The drying of the instrument is sometimes the key to the success of the experiment. The easiest way to dry glassware is to stand overnight. Flasks and beakers can be stored upside down on a drying rack to dry in air slowly. Drying oven can be used to dry glassware. The oven temperature should be set at 110-120 ℃ for several hours before drying. It should be noted that it is necessary to wait until the oven is cold to room temperature before the glassware can be taken out. Otherwise, the water vapor will condense in the wall of the glassware. Rapid drying can be achieved by rinsing the glassware with small amounts of organic solvent such as acetone or 95% ethanol. When using this method, let the water in the glassware is thoroughly run out. It is then rinsed with one or two small portions (about 10 mL) of organic solvent, which is a cheaper grade of solvent, and reagent-grade solvent should not be used. The used solvent should be returned to recycle container. Then use a hair drier to evaporate the solvent thoroughly.

1.6 Heating and Cooling

1.6.1 Heating

Heating is an important laboratory technique, and it serves a variety of functions. It increases the rate of chemical reactions and is used in the distillation of liquids and in the dissolution of solids during the course of recrystallization and purification. There are several common heat sources such as gas, alcohol and electric power in the organic chemistry lab. Heating method can be divided into direct and indirect ones. To avoid possible problems from direct heating, use the following alternate heating methods when necessary.

1.6.1.1 Air bath

With the principle of indirect heating by hot air, the boiling point of the liquid should be above 80 ℃. The simplest air heating is on an asbestos network between open fire and heating objects, but this method usually causes non-uniform heating, which cannot be used for low boiling and flammable liquids reflux or vacuum distillation, etc. The most common air bath is an electric heating mantle. It can be heated to 300 ℃. It is often used for reflux, but the heating temperature is not easy to control, therefore it is not suitable for distillation or vacuum distillation. To avoid overheating, the reaction flask should be kept 1-2 cm away from the inner wall of the heating mantle. One should avoid spilling chemicals into the heating mantle.

1.6.1.2 Water bath

Generally, a heated water bath is a convenient way to heat a liquid below 80 ℃. However, it should be noted that the vessel cannot touch the bottom of the water bath and must be immersed into a water bath so that the surface of the water bath is kept higher than inside surface of solution. And in the operation using potassium and sodium, do not use water bath.

1.6.1.3 Oil bath

For temperature in the range of 100-250 ℃, an oil bath is generally used. The advantage of oil bath is that the temperature is easy to control and the reactants in the container are heated evenly. The commonly used bath liquids are as follows.

1. Glycerol can be heated to 140-150 ℃, if the temperature is too high, it will decompose and release an unpleasant odor. Due to its high hydrophilicity, Glycerol should be heated to remove the absorbed moisture before use after laying aside for a long time.

2. Vegetable oil such as rapeseed oil, Castor oil and peanut oil, etc., can be heated to 220 ℃, and 1% of p-diphenol is often added as an antioxidant to improve the thermal stability of the oil. It should be noted that plant oil will decompose at high temperature and will burn when it reaches the flash point.

3. Liquid paraffin can be heated to about 200 ℃, it is flammable at high temperature.

4. Silicone oil is an ideal bath liquid, but quite expensive for general use. It can be heated to 250 ℃, with good transparency.

Mineral oil is flammable. Care must be taken not to spill any on hot heater. In addition, if any water gets into a mineral-oil heating bath, there is the danger of hot oil spattering out when the temperature exceeds 100 ℃ and the denser water begins to boil. Flasks, when removed from the oil bath, should be allowed to drain for several minutes above the bath and then wiped with paper or rag.

Pay attention to the safety when heating an oil-bath to prevent from accidental fires.

1.6.1.4 Sand bath

Sand bath can be used for heating liquids with boiling points above 80 ℃, especially suitable for those with heating temperatures above 220 ℃. When making a sand bath, an iron plate is filled with sand. The reaction vessel is then partially covered by sand. The sand bath is not used widely now because its temperature rises at a slow rate and is not easy to be controlled.

1.6.1.5 Microwave reactors

In the 1990s microwave reactors designed for laboratory use became commercially available and now microwave heating, or microwave-assisted organic synthesis (MAOS), is routine in many laboratories. Compared to conventional heating methods, modern microwave devices typically allow reactions to proceed safely at much faster rates, using less energy, and often with higher yields and fewer side products. The most important mechanisms by which microwaves heat organic materials are dipolar polarization and ionic conduction. Laboratory microwave reactors are either multimode or single-mode units. Multimode reactors look similar to kitchen microwave ovens.

Note: Household microwave ovens must never be used to heat chemical reactions. These devices lack the ability to monitor and control reaction temperatures and pressures, and their use can result in fires and explosions. Only scientific microwave reactors should be used in the laboratory, and only in strict accordance with the manufacturers' instructions.

1.6.2 Cooling

Cooling is needed when the organic experiment has to be carried out at low temperature, such as diazo-reaction. Sometimes the recrystallization also needs low temperature to accelerate the precipitation of crystals. Organic compounds with low boiling points also need cooling method to reduce its volatilization. Several common cooling methods are described below.

1.6.2.1 Ice-water mixture

Water is the most common choice for its low price and high heat capacity. The common cooling bath is a slush of crushed ice and water. Because of the inversion of density of water at 4 ℃, an ice bath should be stirred well if it is desired to maintain the whole bath at 0 ℃.

1.6.2.2 Ice-salt mixture

Temperature below 0 ℃ can be obtained by mixing inorganic salts with ice or cold water. Table 1.6.1 lists the proportions of ingredients to be mixed to obtain the stated temperature.

Table 1.6.1 Ice-salt system and cooling temperatures provided

Salt	Mass ratio (salt ∶ ice)	Lowest temperature/℃
NH_4Cl	1 ∶ 4	−15
NaCl	1 ∶ 3.3	−21
$NaNO_3$	1 ∶ 2	−18
$CaCl_2 \cdot 6H_2O$	1 ∶ 1	−29
$CaCl_2 \cdot 6H_2O$	1 ∶ 0.7	−55

1.6.2.3 Dry ice or the mixture of dry ice and organic solvent

Dry ice (solid carbon dioxide) can be used to achieve a temperature below −60 ℃. If mixing the dry ice with methanol, isopropanol, acetone, etc. a temperature down to −78-−55 ℃ can be generated. It should be noted that fierce foaming occurs as solid carbon dioxide chunks are added to these organic solvents.

1.6.2.4 Liquid nitrogen

Liquid nitrogen can be used to make a cooling temperature of −195.8 ℃.

Liquid nitrogen and dry ice should be stored in Dewar flasks (wide mouth thermos) or other well insulated containers to maintain their cooling effect. Always use eye protection when using a Dewar flask.

1.6.2.5 Low-temperature bath

Low-temperature bath is a mini refrigerator with an upward opening door assembly, in which a cylindrical stainless-steel tank filled with alcohol is used as the evaporator. A circulation pump is always installed outside the low-temperature bath for pumping the cold alcohol into a condenser. A thermometer or some other indicators can be installed in a low-temperature bath. The reaction flask is immersed in alcohol liquid. It is suitable for the reaction in the range of −30-30 ℃.

Note that the mercury thermometer cannot be used if the temperature is below −38 ℃, because the freezing point of mercury is −38.9 ℃. Instead low temperature thermometer based on organic solvent (ethanol, toluene, n-pentane) with a little pigment should be used.

1.7 Drying of Solids and Liquids

Drying of solids and liquids

1.8 Requirements of Preview, Record and Report of Experiments

1.8.1 Preview of experiments

Every student should prepare a standardized lab notebook and must carefully and fully preview each experiment before start. The preview notes include the following:

1. Purpose and requirements of the experiment.

2. Experimental principles, including reaction formula (main reaction, and main side reaction) and reaction mechanism, etc.

3. The physical constants of main reactants, reagents and products (check the manual or dictionary) in the experiment, dosage (g, mL, mol, mmol) and specifications, and experimental equipment.

4. Draw the main reaction device diagram, write out the brief steps of the experiment and the operating principle.

5. When doing the synthesis experiment, draw the flow chart of crude product purification.

6. Write down protective measures and solutions for possible problems in the

experiment, especially safety problems (including waste disposal).

1.8.2 Experimental record

An important part of any laboratory experience is learning to maintain complete records of every experiment undertaken and every item of data obtained. Therefore, you must develop the habit of making a record while doing the experiment. Record all observations: color changes, temperature rises, explosions, anything that occurs. List the important chemicals you'll use during each reaction including useful physical properties: the name of the compound, molecular formula, molecular weight, melting point, boiling point, density, and so on. You should have entries for the number of moles and notes on handling precaution. The format of the experimental record generally includes the following:

1. Experiment title.
2. Date.
3. Temperature and air humidity.
4. Procedure.
5. Observation.
6. Notes.

The results can be summarized as follows:

1. Product name:
2. The physical properties of the product:
3. Yield: g   Yield: %

If you have finished the experiment, ask your teacher to sign on your lab record for leaving.

1.8.3 Laboratory report

The experiment report is the summary, analysis and discussion of the experimental process after the experiment. It is required to be true and reliable, with complete data and concise text. It should be clearly organized and neatly written. The report generally covers the following sections:

1. Experiment title and date.
2. Objective and requirement: The reason for conducting the experiment is briefly stated in your own words.
3. Experimental principle (reaction, the main side effects, mechanism etc.).

4. Apparatus and reagents.

5. Main device map.

6. Procedure and observation.

7. Results (e.g. R_f value, melting point, yield, and yield calculation, etc.).

8. Discussion.

This is the most important part of the experimental report. In this part, you need to analyze and interpret the experimental results, give comment on some of the errors in the experiment. You can also put forward suggestions for improvement in the experiment. For example, you can discuss whether you have achieved the expected experimental goal or not, and how to improve it if the goal is not achieved.

Questions

1. Please list the safety features you observed in the laboratory.

2. Locate the first aid kit in your lab and check the medical supplies inside. What should we do if someone's eyes are burned by alkali during the experiment?

Further Reading

New technology and method of organic chemistry experiment

1.9 Literature on Organic Chemistry

Literature on organic chemistry

Chapter 2 Basic Experimental Techniques

2.1 Determination of Melting Point

2.1.1 Method for determination of melting point

Determination of melting point (operation video)

Objectives

1. Define the principle and significance of determining the melting point.

2. Learn the technique of determining the melting point.

Principles

Melting point is defined as the temperature at which the solid and the liquid phase are in equilibrium. A pure crystalline solid usually possesses a fixed melting point. Melting point is the temperature at which the first crystal just starts to melt until the temperature at which the last crystal just disappears. Thus, the melting point (abbreviated m.p.) is actually a melting range. The melting point range of a very pure solid organic compound will be narrow (0.5-1.0 ℃).

In Figure 2.1.1, the change from solid to liquid is shown as a sharp "break" or change in slope of the curve at the melting point. When this temperature is reached, heat is absorbed in breaking down the crystal lattice of the solid. If equilibrium between the solid and liquid is maintained, the temperature does not rise until all of the solid are melted. In the determination of an actual melting point, the temperature increases somewhat in the time required for complete melting. Therefore, the observed melting point is usually a range of at least 1 ℃. Impurities even when present in a small amount, usually depress the melting point and widen the melting point range.

The melting point is an important physical constant of crystalline substance. It is often used to identify organic compounds or to check the purity of the compound.

The methods for determination of melting point include capillary method and digital melting point apparatus.

Method 1:Capillary-tube Method

Apparatus and Reagents

Apparatus:Capillary tube, Thermometer (150 ℃), Rubber ring, Thiele tube, Glass tube (1 cm×50 cm), Rubber ring and iron frame.

Reagents:Liquid paraffin, Urea, Benzoic acid, Mixture of urea and Benzoic acid, Unknown compounds A and B.

Procedures

1. Seal the capillary tube at one end

Light a burner and slowly touch the end of the tube to the side of the flame, rotate it slowly.

2. Load the capillary tube with sample

Jab the open end of the tube into a pile of the solid to be analyzed. The solid must be dry, otherwise the results will be affected that solvent can act as an impurity and affect the melting range. If the solid is granular, pulverize the solid somewhat before packing.

Invert the capillary tube and gently tap the tube on the benchtop to make the solid to fall to the closed end. Then, drop the capillary tube closed side down several times through 30-50 cm length of glass tube. The capillary tube will bounce to hit the benchtop, and the solid will be packed into the bottom of the tube. Failure to pack the solid well may lead to shrink when heating, which may cause confusion as to the correct melting temperature.

If needed, repeat the previous steps to load sample until it is with in a height of 2-3 mm in the tube. It is important that the height of the sample is no higher than 3 mm, otherwise the melting range will be artificially broaden.

3. Assemble the apparatus

Determination of the melting point requires a thermometer and a means of heating the sample. The capillary-tube method is the most commonly used method, which is illustrated in Figure 2.1.2. Take a Thiele tube and clamp it to a ring stand or latticework. The tube is normally filled with clear mineral oil, but it may be darken from oxidation or spilled compounds. If the oil is quite dark, it should be replaced. The oil should be filled to at least 1 cm higher than the top triangular arm (an appropriate oil level is pointed to in Figure 2.1.2), and the oil will not circulate as needed if the height is too low.

Insert a thermometer into a one-holed rubber stopper with a slit down one side. Attach the capillary sample to the thermometer with a tiny rubber band (as indicated in Figure 2.1.2). These tiny rubber bands are often made by cutting pieces of small rubber tubing.

Keep the capillary tube so that the solid sample is lined up with the middle of the thermometer bulb.

Place the rubber stopper and thermometer assembly into the Thiele tube, adjusting the height so that the sample is in the middle of the tube. The rubber band should be adjusted so it is not submerged in the mineral oil, keeping in mind that the oil may expand somewhat during heating. The thermometer should not touch the glass.

4. Determination of the melting point

Heat the apparatus gently on the side arm of the Thiele tube with a small flame. If available, use Bunsen burner with a back and forth motion.

If the expected melting point of the compound is known, heat at a medium rate to 10-15 ℃ below the expected melting point, then slow the rate of heating such that the temperature increases no more than 1 ℃ every minute (i. e., very slowly). The temperature must be incremental as the melting point is approached, that the system can reach equilibrium, making the thermometer temperature closed to an accurate gauge of the solid's true temperature.

If the expected melting point of the compound is unknown, heat the sample at a medium rate during the entire time and determine an approximate melting point. Repeat the process with a fresh sample after allowing the oil to cool to at least 10-15 ℃ below the previous melting point. Use the recommendations in former prompt to perform a more careful assessment of the melting point.

A fresh sample is necessary for a second trial of the determination of melting point, even if the first sample solidifies after cooling. Differences in crystal structure between the original solid and the previously melted solid could lead to different melting ranges.

Notes

1. Never heat a closed system. Always vent the tube.

2. Don't heat the bath too fast; the thermometer reading will lag behind the actual temperature of the heating fluid.

Special Instructions

1. The heating speed should be controlled very slowly when it is close to the melting point. If the heating rate is too fast, the measured melting point will be affected.

2. The sample must be extremely finely ground to make the sample strong, so that the heating will be even. If there are gaps, it will not be easy to transfer heat, which will affect the results.

3. The bath is generally concentrated sulfuric acid or liquid paraffin. Concentrated sulfuric acid can be heated to 250-270 ℃. Be careful when heating. Do not make the temperature too high to avoid decomposition of sulfuric acid and release SO_3; in addition, prevent sulfuric acid from touching the skin and causing burns. Liquid paraffin can be heated to 200-210 ℃, but its vapor is flammable, so be careful when operating.

4. After the thermometer cooled down, wipe the concentrated sulfuric acid with waste paper before cleaning it with water, otherwise the thermometer will be easy to burst.

Suggested Waste Disposal

After the experiment is completed, the waste melting point tube, rubber band, and waste paper are poured into the designated waste bucket.

Method 2: Digital melting point apparatus

WRR melting point analyzer use electronic technology to achieve temperature control, initial melting and final melting digital display. Platinum resistance with linear correction is used as the detecting element. The melting process of the sample in the capillary is observed by the high-magnification magnifier, which is clear and intuitive. The structure of the WRR digital melting point meter is shown in Figure 2.1.3.

Questions

1. Which factors can influence the precision of determining melting point?

2. There are two bottles of organic compounds, the melting points were determined as 130-131 ℃ and 130-130.5 ℃ respectively. Use a simple method to judge whether the two compounds are the same.

2.1.2 Instructions for WRR digital melting point tester

Instructions for WRR digital melting point tester

2.2 Determination of Boiling Point and Simple Distillation

Objectives

1. Define the significance of determination of boiling point.

2. Learn the operation of simple distillation and the method for determination of boiling point.

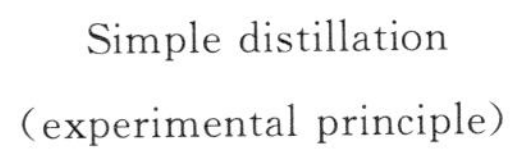

Simple distillation (experimental principle)

Determination of boiling point (operation video)

Principles

Distillation is the process of vaporizing a substance, condensing the vapor, and collecting the condensate in another container. This technique is useful for separating a mixture when the components have different boiling points. And it is the principal method of purifying the liquid. Four basic distillation methods are: simple distillation, vacuum distillation (distillation at reduced pressure), fractional distillation, and steam distillation.

The boiling point is the temperature at which the vapor pressure of a liquid equals the external pressure surrounding the liquid. Each pure organic compound has a fixed boiling point at a certain pressure. The variation range is within 0.5-1 ℃ when a pure substance is distilled. Therefore, the boiling point of a pure organic liquid is one of its characteristic physical properties, just like density, molecular weight, refractive index, and the melting point. It is used to characterize a new organic liquid by comparing one organic liquid with known sample or identify an unknown organic substance.

However, sometimes an organic compound together with other components can form a binary or ternary azeotrope mixture, which also has a definite boiling point. For example, ethanol used in the laboratory or in industry is a constant boiling mixture of 95% ethyl alcohol and 5% water. Thus, it is hard to say that the liquid with a certain boiling point is a pure organic compound.

There are two experimental methods of determining boiling points. When there are large quantities of material, you can use a simple distillation method (see Figure 2.2.1). With smaller amounts of material, you can carry out a micro-determination by the apparatus illustrated in Figure 2.2.2.

Apparatus and Reagents

Apparatus: Distilling flask, Condenser, Distillation head, Adapter, Water bath,

Funnel, Rubber band, Thermometer, Iron stand, Glass tube.

Reagents: Absolute ethyl alcohol (A.R.).

Procedures

1. Distillation method

The apparatus for a simple distillation is illustrated in Figure 2.2.1. The distilling apparatus include three parts: distilling flask part, condenser part, and acceptor part.

Assemble the apparatus according to the principle of "bottom to top" and "left to right". Make sure all taper joints are connected securely. First clamp the flask to an iron stand. Add several boiling stones (porous clay or carborundum) to minimize bumping during distillation, then use a funnel to add 20 mL of absolute ethyl alcohol to the flask (The distilling flask must not be filled to more than two-thirds of its capacity).

Fit the distillation head into the neck of the flask and the thermometer is inserted through the rubber adapter so that the bulb extends to a depth just below the side arm of the distilling head (see Figure 2.2.1). The thermometer bulb must be in the vapor stream, not above it, in order to give a correct reading.

Clamp the condenser to a second iron stand, supporting the weight at the midpoint with the fixed side of the clamp and screwing down lightly with movable side. Adjust the height and angle of the condenser to match the side arm on the distillation head (Never leave the condenser hanging unsupported from the distilling adaptor!).

Place the distillate take-off adapter snugly on the opposite end of the condenser. Place a 10 mL graduated cylinder under the adapter to act as a receiver for the distillate.

Connect the hose attached to the lower end of the condenser to a water supply and place the free end of the other hose securely in the water through. Make sure that all the ground-glass connections and water connections are tight.

2. Determination of boiling point

First heat slowly with a water bath or heating mantle, then gradually increase the power to make the liquid boil, and adjust the heating source to maintain the distillation rate at 1-2 drops per second.

Watch carefully as the layer of warm vapor rises slowly into the distilling adapter and begins to bath the thermometer bulb. Record the boiling point temperature when the first drop of distillate having been collected in the receiving cylinder. This value will be considered as the initial boiling point of the solution. Keep collecting the distillate until not too much solution left. Don't distill to dry. A dry residue may explode on overheating, or the flask may melt or crack.

Finally, turn off the heat source, and then the flow of condenser water when the liquid in the flask cools down. (CH_3CH_2OH b.p. 78 ℃)

3. Microscale boiling point determination

Put 2-3 drops of test solution into a one end closed a glass tube with a diameter of 4-5 mm and a length of 5-6 cm. Put a one end open capillary tube (about 7-8 cm long) with the diameter of about 1 mm in the tube with the sealed upper end out of the solution. After the above-mentioned microscale boiling point tube is installed, fix it on one side of the thermometer with a rubber ring. Keep the bottom of the tube in the middle of the mercury ball. Then add the entire liquid and start heating to make the temperature increase uniformly.

When the boiling point of the sample is reached, small bubbles can be observed continuously at the open end of the inner tube. Once a large number of rapid and continuous bubbles appear, the air inside the spindle has been completely removed. When the vapor of the test solution is replaced, stop heating. The bubbles gradually decrease with the temperature decreases. Write down the temperature when the last bubble retracts back to the capillary, which is the temperature at which the vapor pressure of the liquid and atmospheric pressure balance, that is, the boiling point of the test solution.

Results

The results are recorded in Table 2.2.1.

Table 2.2.1 Data record

Distillate (Volume)	First drop	2 mL	4 mL	6 mL	8 mL	10 mL
Boiling Point/℃						

Notes

1. The distilling apparatus from left to right must be a linear.

2. Please put a few zeolites and open the running water before heating, never forget!

Special Instructions

1. Simple distillation equipment cannot be sealed to avoid explosion due to excessive vapor pressure in the bottle. During atmospheric distillation, the vacuum tail pipe should be open to the atmosphere. If the distillate is toxic or corrosive, it should be introduced into a special device or absorption liquid by a rubber tube; if the distillate is easy to absorb water, install a drying tube on the side pipe of the vacuum tail pipe, and

then communicate with the atmosphere.

2. The upper edge of the mercury bulb of the thermometer is on the same horizontal line as the lower edge of the branch mouth of the distillation flask.

3. A certain distance should be kept between the heat source and the receiver to avoid fire hazards.

4. The condensed water must be passed through before heating, and the condensed water should be "bottom in and top out"; after the experiment, the heat source should be removed first; wait a few minutes, and then stop the water flow after the temperature has cooled slightly.

5. The control of the distillation speed is very important, and it should not be too fast or too slow. During the distillation process, there should always be a stable drop on the mercury bulb of the thermometer, which is a symbol of the balance of gas-liquid equilibrium. At this time, the reading of the thermometer can represent the boiling point of the liquid.

Suggested Waste Disposal

After the experiment is completed, the collected ethanol is poured into a special recycling container for recycling.

Questions

1. In the distillation unit, how does the position of the thermometer affect the temperature reading? What is the best position?

2. Why can't the distilled organics be directly heated by fire?

3. What is the role of zeolite during distillation? Is it possible to add zeolite to a near-boiling solution? Is it necessary to add zeolite again after re-distilling or adding solution?

4. What is the principle of the microcale determination of boiling point?

2.3 Vacuum Distillation

Objectives

1. Understand the principles and applications of vacuum distillation.

2. Master the operation method and installation of vacuum distillation apparatus.

Principles

Vacuum distillation is an important method for separating and purifying organic compounds. It is especially suitable for the substance which has a high boiling point or is unstable under atmospheric pressure. Liquids with boiling point higher than 200 ℃ generally require vacuum distillation separation and purification.

The boiling point of liquid is the temperature at which its vapor pressure is equal to the pressure exerted on it by the atmosphere. Thus, the boiling temperature of a liquid decreases with the decrease of the external pressure. If a vacuum pump is used to reduce the pressure of gas-liquid system, the boiling point of the liquid can be decreased. The operation of distilling at reduced pressure is called vacuum distillation. The corresponding boiling point of the compound to be purified at the selected pressure should be consulted from the literature before vacuum distillation is performed. If such data are not available in the literature, the following rules of thumb can be used to calculate: if the pressure of the system is approximate to atmospheric pressure, for every 1.33 kPa (10 mmHg) decrease in pressure, the boiling point decreases by 0.5 ℃. For a low pressure system, reducing the pressure by one-half lowers the boiling point approximately 10 ℃. A more accurate relationship between pressure and boiling point can be estimated using the nomograph as shown in Figure 2.3.1.

As shown in Figure 2.3.2, the vacuum distillation apparatus is composed of four parts: distillation, vacuum pump, safety protection device and the manometer.

Claisen distillation head is always used in vacuum distillation to lessen the possibility of liquid bumping up into the condenser. The thermometer is inserted into the ground joint with branch pipe, and another ground joint is inserted with a thin capillary tube, which reaches 1-2 mm from the bottom of the flask. During vacuum distillation, the air flows from the capillary into the liquid to produce stable bubbles to promote even boiling. A rubber tube with a screw clamp is arranged at the top of the capillary tube to adjust the amount of air entering the flask to control bumping.

Distilling flask is commonly used as a receiver in vacuum distillation apparatus because of its resistance to external pressure. Flat-bottom flask and conical flask is strictly forbidden in vacuum distillation. A trident dovetail tube can be used for continuous collection of distillates with different boiling points.

The different boiling points of distillates should be taken into full consideration while choosing heat source and condensing tube. Heating on an asbestos net is not allowed, nor should cold water or oil be added into water or oil bath under decompression. The temperature of the hot bath is about 20-30 ℃ higher than the boiling point of the liquid. For compound with higher boiling point, asbestos cloth should be used to wrap the neck of distilling flask to prevent heat loss.

A mercury manometer is generally used in laboratory to measure the pressure of the decompression system. Figure 2.3.2 shows a closed-end manometer. Its U-shaped tube is filled with mercury. The height of the mercury column in the closed tube decreases while the manometer is connected to the vacuum pump. At this point, the difference between the mercury levels in the two tubes is equal to the pressure of the decompression system.

A surge flask should be connected in front of the pump. The screw clamp of the glass tube on the safety flask is used for regulating pressure. If oil pump is used during the distillation, in order to protect the oil pump, a series of washing cylinders or absorption towers should be assembled between the surge flask and the oil pump to absorb water vapor and other gases that may corrode the pump. If a harsh vacuum requirement is unnecessary, water pump will be more convenient.

The vacuum distillation system must be air-tight. Thus, it is usually connected with pressure-resistant silicone tubes. The size of silicone tube should be matched with the hole on the plug. The outer layer of the plug can be coated with collodion to seal the joint.

Apparatus and Reagents

Apparatus: Vacuum distillation apparatus, Oil pump or water pump, Acid and alkali absorber, Iron stand.

Reagents: Ethyl acetoacetate.

Procedures

Place 20 mL ethyl acetoacetate in a Claisen distillation flask. Install the experimental apparatus as shown in Figure 2. 3. 2. Tighten the screw clamp of the capillary. Open the piston of the surge flask. Then, turn on the pump to extract air. Tighten the piston gradually and rotate it to adjust the air flow to the desired vacuum. All connection parts should be checked carefully to prevent air leakage from reaching the required vacuum degree.

After the apparatus has been checked to meet the requirements, the distillation can be started. Turn on the condensate and heat the water bath to boil. At least 2/3 of the main part of the Claisen distillation flask should be immersed in water bath so that 1-2 drops of distillate can be evaporated per second. During the experiment, attention should be paid to the distillation condition and data recording, such as pressure and boiling point.

When distillation is completed, stop heating and remove water bath firstly. When it is slightly cold, open the piston and the screw clamp slowly to balance the internal and external pressure. At last, turn off the oil pump.

Results

The results are recorded in Table 2.3.1.

Table 2.3.1 Data record

Distillate (volume)	First drop	5 mL	10 mL	15 mL
Boiling point/℃				
Vacuum degree/Pa				
Vacuum degree/mmHg				

Notes

1. In case of burst, non-pressure flask and cracked or thin-walled glassware cannot be used in the vacuum distillation system.

2. To control the pressure stability during vacuum distillation, the pump should be started before heating.

3. The piston of the safety flask must be opened slowly after distillation. Otherwise, the mercury column may be broken due to sudden change in pressure.

Special Instructions

1. To prevent explosion caused by large pressure difference between internal and external pressure, distilling flask and receiving flask are required to use pressure-resistant round bottom thick wall glassware.

2. When installing the experimental apparatus, collodion should be applied at the grinding mouth. Checking the tightness of the experimental system before distillation.

3. The liquid can be steamed at low temperature in vacuum distillation. Therefore, do not heat too fast.

4. After the distillation, close and remove the heat source firstly. Then, open the vent valve slowly. Close the vacuum pump when the reading of the manometer returns to zero.

Suggested Waste Disposal

After the experiment, the liquid in the distilling flask and receiving flask should be poured into the organic solvent recovery bucket.

Questions

1. What is vacuum distillation? What compounds can be purified by vacuum distillation?

2. To achieve the maximum vacuum of the decompression system, what problems should be paid attention to?

2.4 Steam Distillation

Objectives

1. Learn the principles and applications of steam distillation.

2. Master the equipment and operation method of steam distillation.

Principles

Steam distillation is a separation process used to purify or isolate volatile organic compounds from nonvolatile compounds such as inorganic salts or the leaves and seeds of plants. Steam or water is added to the distillation apparatus to lower the boiling points of the compounds. When a mixture of two immiscible liquids (e.g., water and organics) is heated and agitated, the surface of each liquid exerts its own vapor pressure as though the other component of the mixture was absent. Thus, the vapor pressure of the system increases as a function of temperature beyond what it would be if only one of the components was present. When the sum of the vapor pressures exceeds atmospheric pressure, boiling begins. Because the temperature of boiling is reduced, damage to heat-sensitive components is minimized.

1. A substance which can be purified by steam distillation should possess the following characteristics:

(1) Insoluble or poorly soluble in water;

(2) No chemical reaction occurs with water in azeotrope;

(3) At about 100 ℃, it must have a certain vapor pressure of 0.067-0.133 kPa (5-10 mmHg), and can volatilize with water vapor.

2. Steam distillation is commonly used in the following situations:

(1) The mixture contains a great amount of resin-like or non-volatile impurities, which are difficult to be separated by distillation, filtration and extraction.

(2) Some high-boiling point compounds which will decompose, deteriorate and discolor during atmospheric distillation.

(3) Separate adsorbed liquid from the mixture containing large portion of solid reactants.

A simple apparatus in common use for steam distillation is shown in Figure 2.4.1. The vapor generator is advisable to hold 3/4 of its volume water. The safety glass tube should be inserted into the bottom of the generator to regulate the internal pressure. If water is gushing from the top of the glass tube, the whole system must be checked to see whether it is blocked (usually the lower mouth of the airway tube in the round bottom flask is blocked).

A T-shaped tube shall be installed between the steam generator and the round bottom flask. A screw clamp is connected at the lower end of the T-shaped tube to remove the water droplets from condensation. The distance between the steam generator and the round bottom flask should be minimized to reduce the condensation of water vapor. The steam generator is heated until it is nearly boiling before the screw is clamped so that the water vapor flows evenly into the round bottom flask during distillation. If the distillation needs to be stopped, it is necessary to open the screw clamp to connect the atmosphere before stopping heating. Otherwise, liquid in the flask will be sucked back into the vapor generator. During the distillation, a rapid water rise in the safety glass tube indicates a blockage in the system. Herein, the screw clamp must be opened immediately, and then remove the heat source. Distillation should be continued after the blockage is resolved.

Apparatus and Reagents

Apparatus: Water vapor generator, Distillation apparatus, Glass bent pipe, Screw clamp, Triangular flask.

Reagents: Naphthalene.

Procedures

Add 4.0 g naphthalene to a 50 mL round bottom flask. Connect the experimental apparatus to vapor generator which filled with water.

Heat the vapor generator until the water boils. Then, close the screw clamp of the T-shaped tube immediately so that the steam pour into the flask. Herein, it can be seen that the mixture is churning in the flask. A mixture of organic substance and water soon appears in the condensing tube. Adjust the flame to prevent the mixture in the flask from boiling over. Control the rate of distillate to 2-3 drops per second. The distillation can be stopped when the distillate is clear and transparent without oil beads. The screw clamp of T-shaped tube should be opened before removing the heat source. Finally, filter and dry the product, and measure the melting point.

Notes

1. The amount of water in the steam generator should not exceed 3/4 of its volume. Insert a glass tube in the bottle. Add water into the bottle when necessary during

distillation.

2. The airway tube must be inserted into the bottom of the flask. To prevent large amount of water vapor from condensing in the flask, it should be heated during the distillation.

Special Instructions

1. The distillate must be cooled before filtration operation after the distillation.

2. If there is solid precipitation appeared in the condensing pipe, the condensed water can be suspended to make the solid melt and flow into the receiving bottle.

3. The screw clamp should be opened immediately when suction occurs.

Suggested Waste Disposal

The product should be poured into recovery container after the experiment.

Questions

1. What is the principle of steam distillation? What is the scope of application?

2. Why the end of the steam tube should be inserted near the bottom of the vessel?

3. After the experiment, what will happen if the heat source of the steam generator is removed first?

2.5 Determination of the Refractive Index

Objectives

1. Define the principle and significance of determination of the refractive index.

2. Master the method of determination of the refractive index with Abbé refractometer.

Determination of the refractive index(operation video)

Principles

The velocity of a ray of light in air is 2.998×10^8 m/s, but is less in transparent liquids of greater density. As a ray of light passes from air to a denser medium, it decreases in velocity. This causes the ray to be bent toward the perpendicular (Figure 2.5.1) at the point of entry of the ray of light, that is, from the point where it intersects the surface of the liquid. The ratio of the velocity of light in air (V_{air}) is defined as refractive index, or the index of refraction(n) of the liquid :

$$n=\frac{V_{vacuum}}{V_{liquid}}=\frac{\sin\alpha}{\sin\beta}$$

Wherein, n represents the refractive index at a specified centigrade temperature and wavelength of light; α represents the angle of the incidence of the beam of light striking the surface of the liquid; β represents the angle of refraction of the light beam in the medium (Figure 2.5.2).

The calculated refractive index depends on the temperature of the liquid and the wavelength of the light used. In normal practice, the refractive index is recorded using the D line of the sodium spectrum (wavelength=589 nm^{-1}) at a temperature of 20 ℃. Under these conditions, the refractive index is reported in the following form:

$$n_{D}^{20}=1.489\ 2$$

The superscript represents the temperature, and the subscript indicates that the sodium D line was used for the measurement.

If the measurement is made at other than 20 ℃, a temperature correction must be applied. The refractive index can be determined to an accuracy of approximately one part per 10 000 for the pure liquid and is normally recorded using four significant figures. Since the value can be determined with such great accuracy, it is of greater precision than melting points, boiling points and similar physical constants as a means of identification of a pure liquid. In practice, it is very difficult to obtain a liquid sufficiently pure to give the values recorded in the handbook for various compounds.

The instrument used to measure the refractive index is called a refractometer. The most common instrument is the WAY-2S Abbé refractometer (Figure 2.5.3).

Apparatus and Reagents

Apparatus: WAY-2S(3S) Abbé refractometer.

Reagents: 95% Alcohol, Distillation water, Unknown samples of A, B, C, D.

Procedures

1. Press "POWER" button 4.

2. Open the refracting prisms unit 11, check the surfaces of the upper and lower prisms, and carefully clean their surfaces with 95% alcohol and lens paper.

3. Place the sample on the lower prism. For the liquid sample, take a few drops of the liquid on the lower prism. Then swing the upper prism back over the lower one and gently close the prisms.

4. Raise the light on the end of the movable arm 10 so that the light illuminates the upper prism.

5. Observe the field of view through eyepiece 1, rotate the adjustment knob 9 until the dark and light fields are exactly centered on the intersection of the crosshairs in the eyepiece (Figure 2.5.3).

6. Rotate the dispersion correction knob 2 in the notch under the eyepiece sleeve until the interface is sharp and uncolored (achromatic).

7. Press the "read display" button and record the refractive index and then press the "temperature display" button to measure and record the temperature.

8. Open the prism, blot up the sample with lens paper, and follow the cleaning procedure with 95% alcohol. Finally put the lens paper between the two prisms.

Results

The results are recorded in Table 2.5.1.

Table 2.5.1 Data record

Sample	Measured temperature	n_D^t				Sample name
		Practical value	Theoretical value	Corrected value	Difference value	
A						
B						
C						
D						

Notes

1. The instrument can be calibrated with distilled water. The refractive index of distilled water at different temperatures is shown in Table 2.5.2.

Table 2.5.2 The refractive index of distilled water at different temperatures

Temperature/℃	n_D^t	Temperature/℃	n_D^t
18	1.333 16	25	1.332 50
19	1.333 08	26	1.332 39
20	1.332 99	27	1.332 28
21	1.332 89	28	1.332 17
22	1.332 80	29	1.332 05
23	1.332 70	30	1.331 93
24	1.332 60		

2. Before and after using the instrument, or when changing the sample, the working surface of the refracting prisms system must be cleaned and wiped.

3. If the sample is solid, it must have a polished smooth surface. Put 1-2 drops of transparent liquid (such as naphthalene bromide) with a higher refractive index than that of the solid sample on the lower prism, there is no need to cover the upper prism.

4. $n_D^{20} = n_D^t + 0.000\ 45 \times (t - T)$, t represents the actual temperature, $T = 20$ ℃.

Special Instructions

1. The tested sample is not allowed to contain solid impurities. When testing solid samples, the working surface of the prisms should be prevented from being roughened or indented. It is strictly forbidden to test samples with strong corrosiveness.

2. The detection of toxic substances should be carried out in a fume hood.

3. Pay attention to protect the prism when using it, and only use lens paper instead of filter paper when cleaning.

4. Do not touch the prism with tip of the dropper.

Suggested Waste Disposal

After the experiment is completed, the experimental garbage and test drugs are poured into a special recycling container for recycling.

Questions

1. A compound has a refractive index of 1.3968 at 17.5 ℃, calculate its refractive index at 20.0 ℃.

2. Write a stepwise summary of using an Abbé refractometer.

2.6 Determination of Optical Rotation

Determination of optical rotation

2.7 Recrystallization and Filtration

Recrystallization and filtration

2.8 Liquid-Liquid Extraction

Objectives

1. Understand the principles of liquid-liquid extraction.
2. Command the operation method of separating funnel.

Principles

Extraction is one of the most important techniques used in the separation and purification of organic compounds. Its basic principle is to make use of the difference in solubility or partition coefficient of solute in two mutually insoluble solvents, so that the solute is transferred from one solvent to another solvent. After repeated extraction, most of the solute can be extracted, so as to achieve the purpose of separation and purification.

Liquid-liquid extraction, on one hand, can be used to extract the desired substance from the liquid mixture, and on the other hand can be employed to remove impurities in the mixture.

The main theoretical basis of extraction method is based on the distribution law. When a solute is dissolved in two immiscible liquids that are in contact with each other at certain temperature, the concentration ratio of the solute in the two liquids is a constant, which is named the distribution coefficient. The distribution coefficient can be expressed as following formula:

$$K=\frac{c_A}{c_B}$$

Where, c_A, c_B represents the concentration of solute in liquid A, B respectively.

The amount of solute remaining after repeated extraction can be calculated by using the distribution law.

Suppose that liquid A (V mL) contains solute W_0. If L mL of liquid B is used for the first extraction, the mass of solute remaining in liquid A equals W_1.

$$K=\frac{c_A}{c_B}=\frac{\frac{W_1}{V}}{\frac{W_0-W_1}{L}}=\frac{W_1L}{V(W_0-W_1)} \quad \text{or} \quad W_1=W_0\frac{KV}{KV+L}$$

If L mL of liquid B is used for the second extraction, the mass of solute remaining in liquid A equals W_2. One could have the following equation:

$$K=\frac{\frac{W_2}{V}}{\frac{W_1-W_2}{L}}=\frac{W_2L}{V(W_1-W_2)} \quad \text{or} \quad W_2=W_1\left(\frac{KV}{KV+L}\right)=W_0\left(\frac{KV}{KV+L}\right)^2$$

Obviously, after n times extraction, the mass of remaining solute in liquid A:

$$W_n=W_0\left(\frac{KV}{KV+L}\right)^n$$

It can be seen that W_n decreased with the increase of n. That means, the result of multiple extraction (dividing a certain amount of solvent into n parts) is better than using the total amount of solvent for only one extraction.

Apparatus and Reagents

Apparatus: Separating funnel (60 mL), Erlenmeyer flask, Beaker (100 mL), Base buret, Iron stand, Volumetric (25 mL).

Reagents: 15% Acetic acid solution, Diethyl ether, Standard sodium hydroxide solution ($0.2\ mol \cdot L^{-1}$), Phenolphthalein indicator, Vaseline.

Procedures

Pour 5 mL of 15% acetic acid solution and 14 mL diethyl ether into a 60 mL separatory funnel. Shake the mixture several times after covering the stopper. The gas inside the separatory funnel should be released at any time to balance the internal pressure caused by the gasification of the vibrating ether. Then, stand the separatory funnel on the iron ring. Remove the stopper of the funnel to make sure the system contact with the atmosphere. When the two liquid phases are completely stratified, open the piston slowly so that the lower liquid flows into an erlenmeyer flask. Add 5 mL water into the erlenmeyer flask. Use phenolphthalein as indicator to titrated the liquid with standard sodium hydroxide solution. Calculate the acetic acid content left in the water. Pour the upper layer of diethyl ether from the separating funnel into the recycling bottle.

As the extraction method described above, 7 mL diethyl ether was used to extract another 5 mL acetic acid solution (15%). Separate the lower aqueous solution and place it in a beaker. Pour the upper layer of diethyl ether into the recycling bottle. Pour the acetic acid solution into the separatory funnel. Then, 7 mL diethyl ether was used to extract it for the second time. Add 5 mL water into the lower layer solution and titrate with standard sodium hydroxide solution. Calculate the amount of acetic acid left in water. The results of one extraction were compared with those of two extractions.

Notes

1. The separatory funnel with a volume 1-2 times larger than the extraction liquid is preferred. Add water to check whether the glass piston and plug are leaking before use.

2. When the system is static stratification, it must be connected to the atmosphere [Figure 2.8.1(b)].

3. The emulsification can be avoided by adding a certain amount of electrolyte (NaCl) during extraction.

Special Instructions

1. Figure 2.8.1(a) shows the vibration method of funnel. The palm of one's right hand against the stopper at the top of the funnel. The neck of the funnel should be held with one's fingers. The left hand controls the piston. Tilt the funnel slightly with the piston up and open it to release the gas generated by the vibration. The funnel must not be pointed at anyone.

2. After static stratification, the upper glass plug of the funnel should be opened before separating operation. Otherwise, the liquid can not flow out. That also will cause two phase remixing.

3. The lower liquid flows out from the piston, and the upper liquid is dumped through the grinding mouth.

4. The composition of the two liquid phases should be correctly judged by the density. If it is difficult to distinguish which one is the target, the two liquids can be retained until the end of the experiment.

Suggested Waste Disposal

The two liquids should be poured into corresponding recycling bucket respectively. The inorganic phase can be neutralized to neutral before recovery.

Questions

1. What's the principle of extraction?

2. How to improve the result of extraction? How to choose the suitable solvent?

3. How to separate incompatible liquids by using the separating funnel?

2.9 Chromatography Techniques

Chromatography technique is an experimental method for separation, purification and identification of organic compounds based on the principle of phase partition. The principle of chromatography is based on the difference in the adsorption or solubility (distribution) of each component of the mixture in a stationary phase. The solution of the mixture moves across a stationary phase and is separated after sufficient running time due to the interactions with the stationary phase. The flowing system is called the mobile phase. The mobile phase can be gas or liquid. A fixed substance is called a stationary phase. The stationary phase can be a solid adsorbent or a liquid (adsorbed on a support agent). According to the force of separation, the chromatographic methods can be divided into adsorption chromatography, partition chromatography, ion exchange chromatography, and size exclusion chromatography. According to operating conditions, it can be divided into thin layer chromatography, column chromatography, paper chromatography, gas chromatography and high pressure liquid chromatography. When the polarity of the mobile phase is less than the polarity of the stationary phase, chromatographic method is called normal phase chromatography. When the polarity of the mobile phase is greater than the stationary phase, it is called reversed phase chromatography.

2.9.1 Column chromatography

Objectives

1. Define the significance and principle of column chromatography.

2. Master the general technique of column chromatography.

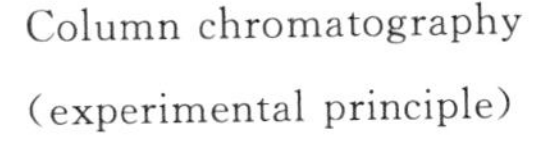

Column chromatography (experimental principle)

Column chromatography (operation video)

Principles

Column chromatography is a kind of chromatographic method. According to the force of separation, it can be divided into adsorption column chromatography, distribution column chromatography and ion exchange column chromatography. The adsorption column chromatography based on the difference in the adsorption of each component of the mixture in a stationary phase is the most widely used.

In the column chromatography, the activated porous substances or powdered solids with a large surface area filled into a glass tube are used as adsorbents, which is the stationary phase. The mixed sample to be separated is introduced onto the top of a cylindrical glass column and then separated with eluent. The adsorbent is then continuously washed by a flow of solvent (moving phase) passing through the column. Since the adsorption capacity of each component in the compound is different, those components move down at different speeds to form several color bands. If the solvent is continued to be eluted, the component with the weakest adsorption capacity will be eluted first (Figure 2.9.1). Throughout the chromatographic process, repeated adsorption-desorption-re-adsorption-re-desorption are carried out to separate the mixture. Collect each component separately, and then identify one by one.

The adsorbent may be almost any material that does not dissolve in the associated liquid phase; those solids most commonly used are alumina, silica gel, magnesium oxide, calcium carbonate and activated carbon. The adsorbent generally needs to be purified and activated, and the particle size should be uniform. The separation effect of column chromatography is related to the particle size of the adsorbent. Generally, the particle size of the adsorbent is preferably 100-150 mesh. If the particles are too coarse, the solution will flow out too quickly, resulting in a poor separation effect. If the particles are too fine, the surface area is large, and the adsorption capacity is high, but the solution flow rate is too slow. There are three types of alumina used in column chromatography: acidic, neutral and basic. The activity of the adsorbent depends on the water content, the most active adsorbent contains the least amount of water. The activity of alumina is divided into I-V levels. The adsorption effect of grade I is too strong and the separation speed is too slow, and the adsorption effect of grade V is the weakest.

In general, the adsorptivity of compounds is proportional to their polarity. The nonpolar compounds pass through the column faster than polar compounds, since they have smaller affinities for the adsorbent. If the adsorbent chosen binds all the solute molecules (both polar and nonpolar) strongly, they will not move down the column. On the other hand, if too polar solvent is chosen, all the solutes (polar and nonpolar) may simply be washed through the column, with no separation taking place. Therefore, the equilibrium competition for the solute molecules between the adsorbent and the solvent should be considered carefully in the column chromatography. The adsorptivity of alumina to various compounds decreases in the following order: Acids or alkalis > alcohols, amines, mercaptans > esters, aldehydes, ketones > aromatic compounds > halogenated substances, ethers > alkenes > saturated hydrocarbons.

The choice of solvent to dissolve the sample is also important. It is usually considered based on the polarity, solubility and activity of the adsorbent of the various components in the separated compound.

The eluent is the solvent used to elute the separated substance from the adsorbent.

The polarity and the solubility of each component of the separated substance are very important for the separation effect. It can be a single solvent or a mixed solution. Generally, solvents with greater polarity are easy to elute the sample, but in order to achieve the purpose of separation, a series of solvents with increasing polarity are generally used.

Initially the components of the mixture adsorb onto the alumina particles at the top of the column. The continuous flow of solvent through the column elutes, or washes, the solutes off the alumina and sweeps them down the column. The solutes (or materials to be separated) are called eluates or elutants; and the solvents, eluents. As the solutes pass down the column to fresh alumina, new equilibria are established between the adsorbent, the solutes, and the solvent. The constant equilibration means that different compounds will move down the column at differing rates depending on their relative affinity for the adsorbent on one hand for the solvent on the other. As the components of the mixture are separated, they begin to form moving bands (or zones), each band containing a single component. If the column is long enough and the various other parameters (column diameter, adsorbent, solvent, and rate of flow) are correctly chosen, the bands separate from one another, leaving gaps of pure solvent in between. As each band (solvent and solute) passes out the bottom of the column, it can be collected completely before the next band arrives.

In this experiment, column chromatography is used to separate the mixture of fluorescein and methylene blue. As neutral alumina is a polar adsorbent, polar compounds (such as fluorescein) adsorb more tightly to the surface of it. Therefore, methylene blue will be washed out of the column first with 95% ethanol as eluent. The fluorescein will then be washed out with more polar eluent.

Apparatus and Reagents

Apparatus: Chromatography column, 150 mL Erlenmeyer flask, 60 mL Separatory funnel, Glass funnel, Pasteur pipette, Filter paper, Cotton, 250 mL Beaker, 100 mL and 25 mL Graduated cylinder. 100 mL and 50 mL Volumetric flask, Balance, Ring stand, Glass rod, Rubber band.

Reagents: Neutral alumina (100 mesh), 0.1 g/L Mixed ethanol solution of fluorescein and methylene blue, 95% Ethanol, Distilled water, Sea sand.

Procedures

1. Preparing the column

The column is packed in two ways: the slurry method and the dry pack method.

Here, we choose the latter one. Clamp the glass chromatography tube in a vertical position onto a ring stand and fill with about 15 mL 95% ethanol. A loose plug of cotton is tamped down into the bottom of the column with a long glass rod until all the bubbles are forced out. Do not use too much cotton, and do not pack it too tightly. Slowly put sea sand into the column until there is a 0.5-1 cm layer of sea sand over the cotton. Any sand adhering to the side of the column is washed down with a small quantity of solvent. The sand forms a foundation that supports the column of adsorbent and prevents it from washing through the stopcock. Add 10 g alumina (adsorbent) to the beaker and then add 95% ethanol with swirling to form thick but flowing slurry. The slurry should be swirled until it is homogeneous and relatively free of entrapped air bubbles. The stopcock is opened to allow solvent to drain slowly into a large beaker at the rate about 1 drop per second. Alternately, the slurry is mixed by swirling and is then poured in portions into the top of the draining column (a wide-necked funnel may be useful here). The column is tapped constantly and gently on the side, during the pouring operation, with a glass rod fitted with a rubber band. The tapping promotes even settling and mixing and gives an evenly packed column free of air bubbles. Tapping is continued until all the material has settled, showing a well-defined level at the top of the column. After all of the adsorbent has been added, carefully pour approximately 0.5 cm of sea sand on top. As shown in Figure 2.9.2, this layer protects the adsorbent from mechanical disturbances when new solvents are poured into the column later. During the entire procedure, keep the level of the solvent above that of any solid material in the column! Check the column, if there are air bubbles or cracks in the column, dismantle the whole business and start over!

2. Applying the sample to the column and elution process

When the solvent level is just at the top of the upper white sand about 1 cm, close the stopcock and add the mixture solution to be separated to the column about 10 drops. After reopening the stopcock and allow the upper level of the solution to reach the top of the sand, fill the column with 95% ethanol solution, and proceed to develop the chromatogram. The methylene blue is washed out until the solvent becomes colorless. Then change water as the elution solvent, and follow the same procedure. Collect the methyl orange in Erlenmeyer flasks.

3. Collecting the eluted compounds

Pure compounds of the mixture are recovered by the evaporation of the solvent in the collected fractions.

Notes

1. Before pack the column, the cotton should be wetted with 95% ethanol solution to prevent air bubbles in the column.

2. The column should be packed evenly and tightly without air bubbles and crackings.

3. In order to maintain the uniformity of the column, the entire adsorbent should be soaked in the eluent or solution during the entire operation. Otherwise, when the eluent or solution in the column dries, dry cracks will occur. The uniformity of infiltration and color development will be affected. Do not disturb the cylinder.

4. The structures of fluorescein and methylene blue are shown in Figure 2.9.3(a).

Special Instructions

In the column chromatography separation, the height of the eluent in the column should never be lower than the top end of the silica gel.

Suggested Waste Disposal

Dispose of any leftover eluent in the recovery container for organic solvent after experiment.

Questions

1. What's the principle of the separation of the two compounds?

2. If there are bubbles or uneven packing in the column, how will it affect the separation effect? How to avoid it?

3. Whether the rate of eluent flow could affect the result of the chromatography? What factors should be considered?

2.9.2 Paper chromatography

Objectives

1. Learn the the principle and application of paper chromatography.

2. Master the method for separation and identification of amino acids by paper chromatography.

Paper chromatography (experimental principle)　Paper chomatography (operation video)

Principles

Paper chromatography is a liquid-liquid partitioning technique. It includes two parts, stationary phase and mobile phase. Filter paper as a support for the stationary liquid, which is usually water. The paper fibers become hydrated when the water molecules form hydrogen bonds to the hydroxyl groups of the glucose units in cellulose. The organic liquid which is partially miscible with water acts as the mobile phase (developing solvent). Compounds are separated according to their relative solubility in two liquid phases. Molecules that are more soluble in water will migrate very slowly on the paper. Molecules that are more soluble in the mobile organic liquid will move fast on the paper. The solvent moves on the paper by capillary action, when it is flowing on the paper and the separation of different molecules is being accomplished.

The relative movement of a compound is usually determined by comparing the distance traveled by the sample with the distance traveled by the solvent. This ratio is called the R_f value, which stands for ratio to front; it is also referred to as retention factor. It is determined by the formula:

$$R_f=\frac{\text{distance traveled by compound (from origin)}}{\text{distance traveled by solvent (from origin)}}=\frac{a}{b}$$

The R_f values depend on the structure of the substance, the solvent system and type of filter paper used. The same substance usually has alike R_f value under the same condition.

Paper chromatography is a powerful tool for the separation of highly polar and /or polyfunctional compounds such as sugars, amino acids and alkaloids. One advantage of paper chromatography is inexpensive; however, its low capacity generally makes it an unsuitable technique for preparative separations. In addition, the long development times observed in many separations are also disadvantageous.

In this experiment, amino acids will be separated on a vertical paper by ascending migration. Since amino acids are colorless, ninhydrin will be applied to the developed chromatogram to locate the different amino acids. Amino acids in an unknown mixture will be identified by a comparison of the R_f values of the unknown amino acids with the R_f values of the known amino acids determined on the same chromatogram.

Apparatus and Reagents

Apparatus: Developing jar (9 cm × 18 cm), Xinhua No. 1 filter paper (6.5 cm × 14 cm), Capillary tube, Blower, Sprayer, Tweezers, Glass plate, Cotton thread, Pencil, Ruler.

Reagents: 0.5% Leucine solution, 0.5% Proline solution, Mixture of amino acids;

Developing agent: supernatant after mixing 1-butanol : glacial acetic acid : ethanol : water= 4 : 1 : 1 : 2, (V/V);

Color developing agent: 0.5% Solution of ninhydrin in ethanol.

Procedures

1. Spotting the filter paper

Take a filter paper approximately 6.5 cm × 14 cm. Touch the filter paper with tweezer or only on one edge. Draw a pencil line across the paper 2 cm away from one edge and parallel to the opposite edge of the paper. Make three marks 1.5 cm apart along this line (Do not get too close to the edge of the filter paper). Label these marks at the top of the paper with "Leu", "Pro" and "unk" to represent leucine, proline and an unknown mixture of two of these amino acids. Dip a length of capillary tube into the leucine solution, lightly touch the filled end of the capillary to the pencil mark on the filter paper marked "Leu" so as to make a spot smaller than 2 mm diameter. Allow the paper to dry, spot the solution on the same place two more times and allow the paper to dry between each application. Repeat the spotting procedure with solutions of proline and the unknown mixture at the appropriately marked locations on the paper. Use different capillaries for each solution.

2. Developing the filter paper

When the sample spots on the paper have dried, hang the paper into the jar by cotton thread. The paper must be vertical and the pencil line must retain above the surface of the liquid with the entire bottom edge immersed in the eluent to a depth of about 2 mm. Cover the jar with top and the development begins. As shown in Figure 2.9.4. After about 60 min, remove the paper from the jar. Mark the solvent front with pencil and dry it with a blower.

3. Visualization

Put the dried paper on the glass plate and spray it with the developing agent(0.5% ninhydrin), dry it and blow hot air from blower over the chromatogram to speed the evaporation of the ninhydrin solvent and the reaction between ninhydrin and the amino acids. The paper treated with ninhydrin produces purple and yellow spots.

4. Calculating the R_f value and identification of samples

Measure the distance from the starting line to the center of each spot and record it. Calculate the R_f value for each amino acid, as shown in Figure 2.9.5. Use the calculated R_f values to identify amino acids in the unknown mixture.

Notes

1. The chromatographic filter paper should have high purity, with appropriate thickness, adsorption and water retention. Generally fast filter paper is suitable for most separation; however, when high resolution is needed, slow filter paper is

preferred.

2. It is better for the amino acid solution to be used once prepared, and it should be stored in the refrigerator.

3. The developing solvent is generally selected according to the properties of the separated components. Selecting a suitable solvent is often a trial-and-error process, particularly if a mixture of solvents is required to give a good separation. A solvent that does not cause any compounds to move from the original spot is not polar enough, whereas a solvent that causes all the spotted sample to move with the solvent front is too polar.

4. The development time depends on many factors, such as the properties of the separated components and the development solvent, the quality of the filter paper and the ambient temperature, etc. For good resolution, reasonable R_f values should be in the range between 0.4 and 0.8.

Special Instructions

1. The filter paper should be clipped with tweezers. For traces of amino acids will be deposited on the paper wherever it is touched. The resulting amino acid deposit is sufficient to react with the ninhydrin reagent used to locate the amino acids examined in this experiment. Therefore, finger prints on the paper anywhere except at the top edge will result in a poor chromatogram.

2. Do not overload the paper! An excessive amount of compound will cause overloading and result in large spots with considerable trailing. The spots from compounds with similar R_f values will overlap, making correct analysis very difficult.

3. Do not permit the solvent to touch the spots.

4. Marks on the paper must be in pencil. Ink from a ballpoint pen will migrate and result in a poor chromatogram.

5. In the experiment, always pay attention to the solvent front, the filter paper should be taken out before the developing agent reaches the upper edge of the filter paper.

6. Cover the chromatographic cylinder lid tightly to ensure the developed solvent saturate the atmosphere above the liquid.

Suggested Waste Disposal

Dispose of any leftover developing solvent in the recovery container for organic solvent after experiment.

Questions

1. What is the principle of paper chromatography? What is the definition of R_f value? What factors will affect the R_f value?

2. Can we predict the order of R_f values for samples with different polar components after paper chromatography in a low polar developing solvent? Please explain it.

2.9.3 Thin layer chromatography

Objectives

1. Define the significance and principle of thin layer chromatography.

2. Master the general technique of thin layer chromatography.

Principles

Thin layer chromatography(TLC) is an important experimental technique for rapid separation and qualitative analysis of mixed substances, and it is also used to follow the course of the reaction.

In TLC, adsorbent of the stationary phase, is evenly spread on a glass plate to form a thin layer. Then the sample is added, and an appropriate solvent is selected as the mobile phase (developing agent). Due to the different adsorption capacity of the mixture components to the adsorbent, the eluent moves at different speeds with each component, so as to achieve the purpose of separation.

Usually the term R_f is used to stand for "ratio to the front" and is expressed as a decimal fraction:

$$R_f = \frac{\text{Distance traveled by compound}}{\text{Distance traveled by developing solvent front}}$$

R_f is a physical characteristic of the compound. For a certain compound under certain conditions, the value of R_f is a constant, and its value is between 0 and 1. The separation and identification of mixed samples can be carried out according to the value of R_f.

There are two types of commonly thin-layer chromatography: adsorption chromatography and partition chromatography. The adsorbents of adsorption chromatography are generally silica gel or alumina, and the support substances of partition chromatography are diatomaceous earth or cellulose.

Silica gel is an amorphous porous substance, slightly acidic, suitable for the separation and analysis of acidic substances. The silica gel used for thin layer chromatography is divided into:

Silica gel H: It does not contain adhesives and other additives;

Silica gel G: It contains calcined gypsum binder;

Silica gel HF_{254}: It contains fluorescent substance, which can observe fluorescence under 254 nm ultraviolet light;

Silica gel GF_{254}: It contains both calcined gypsum and fluorescent agent.

Alumina is similar to silica gel and is also divided into alumina G, alumina GF_{254} and alumina HF_{254}.

Alumina is more polar than silica gel, so it is more suitable for separating less polar compounds (hydrocarbons, ethers, aldehydes, ketones, etc.); on the contrary, silica gel is suitable for separating more polar compounds (carboxylic acids, alcohols, amines, etc.).

The separation and purification experiment of amino acids utilizes the thin-layer chromatography to separate and purify the mixed amino acids. Based on the difference in the adsorption of each component in silica gel, the developing agent migrates up the thin-layer plate, carrying with amino acids at different moving speed. After a certain period of unfolding, they separated from each other. Take out the thin-layer plate, evaporate the solvent, color with ninhydrin, and measure the value of R_f.

Apparatus and Reagents

Apparatus: Chromatography cylinder, Glass plate (10 cm×3 cm), Capillary tube, Blower, Sprayer, Tweezers, Cotton thread, Pencil, Ruler.

Reagents: Silica gel G, 95% Ethanol, 0.1% Arginine, 0.1% Alanine, Mixed amino acids (alanine + arginine), 0.1% Sodium carboxymethyl cellulose, Developing agent (*n*-butanol : acetic acid : Water=12 : 3 : 5), Color developing agent (0.5% ninhydrin acetone solution).

Procedures

1. Preparing the thin-layer plates

Weigh 2.5 g of silica gel G into a small beaker, add 8 mL of 0.1% sodium carboxymethyl cellulose aqueous solution, mix thoroughly, pour it on a clean glass plate, and then tilt and rotate the glass plate to make the support form a uniform thin film on the glass plate. Place the glass plate flat on the desktop, wait for 10-15 min, then place it in an oven at 110-120 ℃ for 30 min to activate it, let it cool, and place it in a desiccator for later use.

2. Spotting

Gently draw a horizontal line with a pencil at a distance of 2 cm from the bottom of the thin layer plate, and then lightly draw 3 points on the line at equal distances. Use a capillary tube to draw alanine, arginine and the mixed solution at the three points. Click on each point 2-3 times (wait for each point to dry before clicking the second time).

3. Developing a plate

Add an appropriate amount of developing agent to a clean chromatographic cylinder, saturate the developing agent in the chromatographic cylinder for about 10 min. And place the end of the chromatography plate with the sample in the chromatographic cylinder obliquely (be careful not to immerse the sample in the developing agent), immediately cover it tightly to expand the sample in a closed chromatography cylinder, as shown in Figure 2.9.6. When the front of the spreading agent reaches 2/3 of the upper part of the thin-layer plate, take out the thin-layer plate, and immediately mark the position of the solvent front, and let the plate dry naturally or with a hair dryer.

4. Visualization

Spray the chromatographic plate with ninhydrin solution and blow dry with an electric hair dryer, and purple spots appear. Measure the distance from the origin to the spot center and the distance from the origin to the front of the solvent, and calculate the R_f value of the sample respectively.

Notes

1. The glass plate used to prepare thin layer chromatography must be flat and clean. Otherwise, the silicone layer is easy to fall off.

2. The activity of the thin-layer plate is related to the water content, and its activity decreases with the increase of the water content. Therefore, after dried at room temperature, the prepared thin-layer plate must be heated and activated in an oven to further remove moisture. The activation conditions should be determined according to the specific situation. Generally, silica gel plates are dried at 105-110 ℃ for 30 min, and alumina plates are dried at 200-220 ℃ for 4 h. When separating some easily adsorbable compounds, it may not be activated.

3. The amount of developing agent added should not exceed the point of the sample on the thin plate.

Special Instructions

1. The silica gel layer should not be damaged when marking with a pencil.

2. Do not overload the thin-layer plate! An excessive amount of compound will cause overloading and result in large spots with considerable trailing. The spots from compounds with similar R_f values will therefore overlap and correct analysis will be very difficult.

3. Do not permit the solvent to touch the spots.

4. Marks on the thin-layer plate must be in pencil. Ink from a ballpoint pen will migrate and result in a poor chromatogram.

5. Cover the chromatographic cylinder lid tightly to ensure the developed solvent saturate the atmosphere above the liquid.

Suggested Waste Disposal

Dispose of any leftover developing solvent in the recovery container for organic solvent after experiment.

Questions

1. What is the principle of thin layer chromatography?

2. During chromatography, what will be the result if the developing agent exceeds the sample spots?

3. Why can the qualitative analysis of the tested sample be carried out based on the value of R_f?

2.10 Paper Electrophoresis of Amino Acids

Objectives

1. Understand the principle of electrophoresis.

2. Master the technique of separation and identification of amino acids by paper electrophoresis under normal voltage.

Paper electrophoresis of amino acids (experimental principle)

Paper electrophoresis of amino acids (operation video)

Principles

Electrophoresis technique appeared slightly later than chromatographic technique. It is a method combined with chromatography, also known as electrophoresis chromatography.

Charged particles, in the electric field, move toward the electrode which has opposite charge to the particle. This is called electrophoresis. Under certain conditions (such as electric field strength, pH value), the migration rate of each component is influenced by its relative molecular mass as well as the nature and quantity of its own charge. Therefore, in the same electric field, their migration directions and distances are also different in a certain time, so that the purpose of separation and identification of some substances can be achieved. The migration rate of charged particles in the electric field is not only related to their own properties, but also affected by factors such as electric field strength, solution pH, ionic strength and electroosmosis.

Electrophoresis techniques can be classified as either moving boundary electrophoresis (solution electrophoresis) or zone electrophoresis. According to the differences of the supporters, the zone electrophoresis can be grouped into: paper electrophoresis, thin layer electrophoresis, cellulose acetate electrophoresis, starch gel electrophoresis, polyacrylamide gel electrophoresis, etc.

Paper electrophoresis can be used for the separation and identification of amino acids, proteins, sugars, organic acids and other substances.

Amino acids, for example, are charged molecules. In their structures, the acidic group-carboxylate anion ($—COO^-$) and the basic group-ammonium cation ($—NH_3^+$) are attached to the same carbon. They are zwitterions (amino acid molecule that has one positive and one negative charge). In acidic solution (with respect to the isoelectric pH of the amino acid), the carboxylate anion picks up a proton and is converted to a carboxyl group ($—COOH$). As a result, the zwitterion is transformed to a cationic form that has a net positive charge. In this case, amino acid moves toward the negative electrode. While in basic solution, the positively charged ammonium ion ($—NH_3^+$) gives up its proton to the hydroxide ion, the zwitterion is converted to a neutral amino group. Thus, amino acid exists as anion form, which moves toward the positive electrode. Because the ionizing constants of acidic or basic group in amino acids are different, each amino acid has its own migration direction and rate in the electric field, by which we can separate and identify all kinds of amino acids, as shown in Figure 2.10.1. In this experiment, the isoelectric points of glutamic acid, alanine and arginine are 3.22, 6.02 and 10.76, respectively. They migrate in different directions and distances under the electric field. The buffer solution of pH 5.9 is used as the supporting medium during electrophoresis.

Apparatus and Reagents

Apparatus: DYY-4C stationary voltage electrophoresis apparatus, Electrophoretic tank, Filter paper (zhonghua No.3, 4 cm×15 cm), Glass plate, Blower, Tweezers,

Sprayer.

Reagents: Mixture of amino acids (Alanine, Arginine and Glutamic acid), Buffer solution(pH = 5.9) of monopotassium o-phthalic acid-sodium hydroxide.

Developing agent: 0.5 % ninhydrin in absolute alcohol.

Procedures

1. Spotting sample and wetting the filter paper

Take a filter paper strip with tweezer. Use a pencil draw a straight line in the middle of the sheet and make one mark in the middle of the line. Then label "+" and "−" at both ends. Dip a length of capillary tube into the mixture solution and lightly touch the filled end of the capillary to the pencil mark on the filter paper so as to make a spot smaller than 2 mm. Dry it with a blower.

2. Running electrophoresis

First pour the buffer solution into electrophoretic tank (Note: The solution level of both sides must be the same). Then pick up the filter paper with tweezers and dip it into the buffer solution, which soon gets wet. Take it out of the solution and arrange it horizontally on the bridge of the electrodes, both ends of the paper dipping into buffer solution. Now you can turn on the direct current after putting the cap of the tank, the electrophoresis begins. First adjust the voltmeter to 100 V, and then increase 10 V every 1 min until the voltage is 220 V. During the process, voltage must be controlled at 220 V (current 10-15 mA). After 60 min you may turn off the direct current to stop electrophoresis. Take the paper out, place it on the glass plate, and dry it with a blower.

3. Color developing

Put the dried paper on the glass plate and spray it with a developing agent (0.5% ninhydrin solution), which reacts with amino acids to give a purple color. Dry it with a blower. The paper treated with ninhydrin produces a series of colored band, known as an electrophoretic pattern. Basing on it, you can determine the position of various amino acids in the paper. If the mixture is unknown, we compare the electrophoretic pattern of it with that of a known for identification, and then you can easily find out what amino acids they are. Electrophoresis is a valuable technique not only for identification and separation but also as a medical diagnostic tool.

Draw out the result of the paper electrophoresis of amino acids as Figure 2.10.2.

Notes

1. The migration rate of a charged particle in the electric field not only depends on

its own properties, but also is related to the following factors.

(1) Electric field strength: It plays an important role in electrophoresis. The higher the electric field strength, the faster the charged particles move. Therefore, according to the strength of electric field, electrophoresis can be classified into normal (10-50 V/cm), low (less than 10 V/cm) and high (greater than 50 V/cm) voltage electrophoresis.

(2) The pH of the solution: The pH of the solution has a greater impact on the charged state of certain molecules such as amino acids, proteins and other ampholytes.

(3) Ionic strength of the solution: the stronger the ionic strength of the solution during electrophoresis, the slower the electrophoresis rate will be; if the ionic strength is too low, the buffer capacity of the buffer solution is so small that is not suitable to maintain a stable pH. Generally suitable ionic strength is 0.02-0.2 mol/kg.

(4) Electroosmosis: Under the action of an electric field, the relative movement of the liquid to the solid support is called electroosmosis. Since electroosmosis often coexists with electrophoresis, it has an impact on the moving distance of charged particles. If the direction of electrophoresis is the same as that of electroosmosis, the speed of electrophoresis will increase; on the contrary, the speed of electrophoresis will slow down. Therefore, try to choose a material with less electroosmotic effect as a support.

(5) Other factors: the viscosity of the buffer solution, the interaction between the buffer solution and the charged particles, and the temperature change during electrophoresis can also affect the migration rate of electrophoresis.

2. The substances separated by paper electrophoresis are limited to charged substances, so the application range is not as wide as that of paper chromatography. However, for some substances, especially macromolecular compounds, paper chromatography is not as simple as electrophoresis, thus paper electrophoresis has become one of the commonly used analytical methods. It is commonly used in the separation and identification of proteins and amino acids in medicine.

3. Preparation of monopotassium o-phthalic acid-sodium hydroxide buffer solution (pH=5.9): Weigh sodium hydroxide 0.86 g and monopotassium o-phthalic acid 5.1 g, Add water to 1000 mL.

Special Instructions

1. In order to avoid the influence of water evaporation on the pH value of the solution, electrophoresis is generally carried out in a closed system.

2. Reproducible results can be obtained only when the voltage is stable, so the

voltage should be gradually increased.

3. Don't touch the paper strip to avoid too many fingerprints after color development, which will affect the observation results.

4. To prevent electric shock, the power must be cut off before the filter paper is taken out, and the filter paper should not be touched with hands or tweezer during the electrophoresis process.

Suggested Waste Disposal

The used electrophoretic solution and filter paper should be put into special recycling containers after the experiment.

Questions

1. What is the isoelectric point of amino acids? What's the correlation between the isoelectric point and the charged state of the amino acid molecule?

2. Write the structural formulas of alanine, arginine, glutamic acid in buffer solution at pH=7, and determine which electrode do they migrate to in electrophoresis.

2.11 Spectroscopic Technology

Spectroscopic technology

2.12 Molecular Model Exercises

Objectives

1. Observe stereoscopic model of organic compound molecules and master the way of drawing the three-dimensional structures of organic molecules.

Molecular model exercises (experimental principle)

2. Master the concept, classification and nomenclature of isomer, structural isomer and stereoisomer. Explore the relationship between the structures and properties of organic compounds.

Principles

Isomerism is very common in organic chemistry. Isomerism can be summarized as Figure 2.12.1.

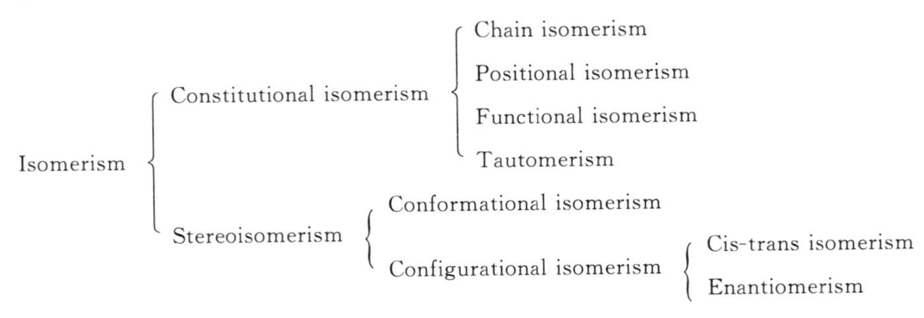

Figure 2.12.1 Common isomerism

Constitutional isomers have the same molecular formula but a different connectivity of atoms in their molecules. Stereoisomers have the same molecular formula and the same connectivity but different orientations of their atoms in space. The free rotation of the single bond is responsible for conformational isomerism. Generally, the interconversion of conformational isomers can be achieved through the rotation of carbon-carbon single bonds with a small barrier at room temperature, which does not involve valence bond cleavage. Pure conformers cannot be isolated in most cases, because the molecules are constantly interchanging through all the possible conformations. In configurational isomers, the positions of the atoms cannot interchange so as to be identical simply by rotations along single bonds and their interconversion would require breaking of the valence bond with high energy. Therefore, configurational isomers can exist stably at room temperature, and can be separated from each other based on their different properties. A molecule that is not superimposable on its mirror image is said to be chiral. Each isomer of the mirror image pair is called an enantiomer.

The structure of a molecule is represented by structural formula, as shown in Figure 2.12.2. The commonly used structural expressions are the straight-line notation (Kekulé structure), condensed formulas and the bond-line formula. The three-dimensional structure of a compound has four types of drawings: wedge perspective structure, sawhorse perspective structure, Newman projection and Fischer projection. For example, two thin solid lines indicate that the two atoms are in the plane of the paper; the hashed-wedged line (❙❙❙❙❙❙) corresponds to the bond that lies below the plane of the paper, and the solid-wedged line (▬) to that lying above the plane of the paper.

Conformational isomers are represented in two ways, as shown in Figure 2.12.3. A sawhorse representation views the carbon-carbon bond axis from an oblique angle of 45° and all the bonds are represented by solid lines. The angle between the other three bonds on each carbon atom is 120°. In a Newman projection, a molecule is viewed down the axis of a C-C bond. The three atoms or groups of atoms on the carbon nearer your eyes are shown on lines extending from the center of the circle at angles of 120°, and the other three atoms or groups of atoms on the carbon farther from your eyes are shown on lines extending from the circumference of the circle, also at angles of 120°.

Fischer projections are now a common means of representing stereochemistry at chirality centers, particularly in carbohydrate chemistry. The basic principles of drawing Fischer projection are: (1) Place the main chain vertically and the carbon atom with the lowest ranking (or the carbon atom with higher oxidation state) at the top; (2) The horizontal lines represent bonds coming out of the page, and the vertical lines represent bonds going into the page; (3) A chiral carbon atom is represented in a Fischer projection by the point at which lines cross. The configuration of *L*-lactic acid can be drawn as Figure 2.12.4.

The cyclic structure of carbohydrate compounds is generally represented by Haworth structure as Figure 2.12.5.

Molecular structure is usually represented by a molecular model. The general used model is "ball-and-stick" model. Different color small balls are used to symbolize the various atoms or groups.

For example:

Carbon (C) —black Hydrogen (H) —white Oxygen(O)—red

Sulfur (S)—yellow Nitrogen(N)—blue Phosphor (P)—purple

Chlorine (Cl)—green

The various balls may be joined together with short stick, casing pipe or spring, indicating chemical bond of atoms or groups.

The carbon atom may be bonded with each other by single, double and triple bonds to form a chain or ring.

Model material: a bag of organic molecular models.

Model Exercises

1. Build up molecular models of methane, ethylene and acetylene to observe and compare single, double, and triple bonds and bond angles. Draw the three-dimensional

structural formula (wedge perspective) of methane, ethylene and acetylene molecules.

2. Build up a molecular model of ethane, propane, *n*-butane and isobutane.

(1) Observe and compare the carbon chain extensions of *n*-butane and isobutane to identify methyl, ethyl, propyl, isopropyl and primary, secondary and tertiary carbon atoms.

(2) Change the *n*-butane and isobutane models into *n*-butyl, isobutyl, *sec*-butyl and *tert*-butyl groups and write the structural formula (straight line) of the above groups.

(3) Rotate the C—C single bond by 360° in the ethane molecular model. Observe its conformation and identify the eclipsed and staggered conformation from these conformations and draw the sawhorse and Newman projection formulas.

(4) Rotate the C2—C3 single bond by 360° in the *n*-butane molecular model. Observe its conformation, and draw the sawhorse and Newman projection to show the anti (staggered), gauche (staggered), eclipsed, totally eclipsed conformational isomers and point out the dominant conformation.

3. Build up models of all isomers of butene, and name 2-butene with *cis*- and *trans*-, or *Z* and *E* configuration nomenclature. Draw two three-dimensional structural formulas (perspective), and observe the relative distance of the two methyl groups.

4. Build up a model of butenedioic acid, determine its *cis*- and *trans*- isomers and draw the three-dimensional structural formula.

5. Build up molecular models of lactic acid, tartaric acid and glucose.

(1) Observe the lactic acid molecular model, draw the wedge perspective formula and Fischer projection formula of the lactic acid molecule, and understand the configuration nomenclature of *D*-lactic acid, *L*-lactic acid, *R*-lactic acid and *S*-lactic acid.

(2) Observe the molecular models of tartaric acid. Draw the Fischer projection formulas of the three stereoisomers of tartaric acid, and find the symmetry factors (such as symmetry plane and symmetry center) of the *meso*-tartaric acid molecule.

(3) Observe the molecular models of the open chain structure and cyclic structure of glucose. Draw the Fischer projection formula of the open-chain structure and the Haworth formula of the cyclic structure of the glucose molecule, understand the transition relationship between the open-chain structure and the cyclic structure, and compare the stability of the two conformations of α-*D*-glucose and β-*D*-glucose.

6. Build up models of all stereoisomers of 2-bromo-3-chlorobutane and draw Fischer

projections for them and assign the absolute configurations (R, S) to all the asymmetric atoms.

7. Try to explain the relationship between the compounds in the following two groups by molecular models (same compound? enantiomer? or diastereomer?), and write out the configurations with R/S nomenclature.

(1)

(a)	(b)	(c)	(d)
top CH_3; H—C—Br; bottom Cl	top CH_3; H—C—Cl; bottom Br	top CH_3; Br—C—H; bottom Cl	top CH_3; Cl—C—H; bottom Br

(2)

(a)	(b)	(c)	(d)
top CH_3; H—C—Br; Br—C—H; bottom CH_3	top CH_3; Br—C—H; H—C—Br; bottom CH_3	top CH_3; H—C—Br; H—C—Br; bottom CH_3	top CH_3; Br—C—H; Br—C—H; bottom CH_3

8. Try to find out which of the following compounds from (a) to (f) have the same configuration with models, and name with R, S nomenclature.

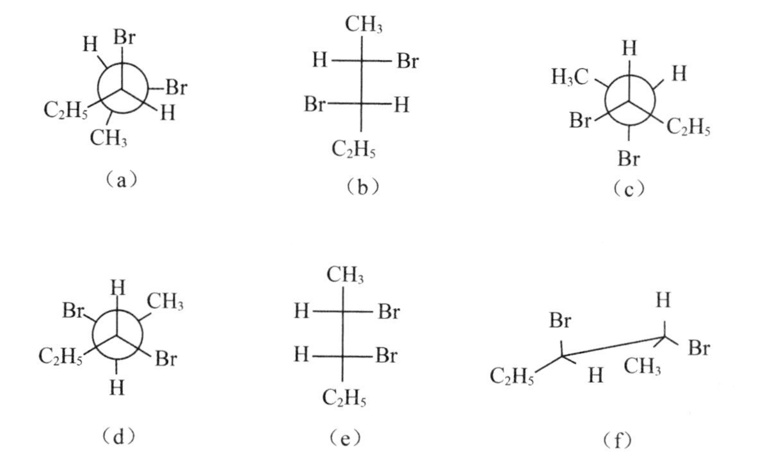

9. Build up molecular models of cyclohexane, compare the stability of the boat conformation and the chair conformation, and draw the sawhorse formula of the chair conformation.

10. The followings are two conformations of methylcyclohexane, try to explain

which one is more stable.

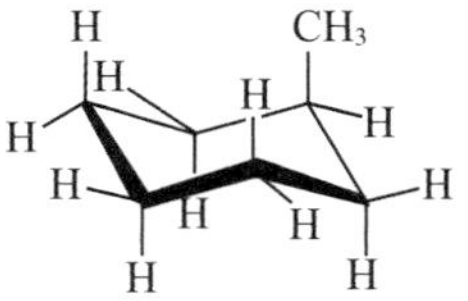

(a) 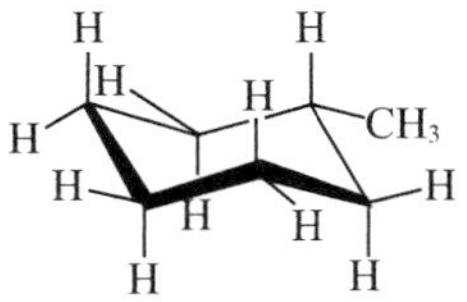　　　　　　　　(b)

11. Build up molecular models of 1,4-dimethylcyclohexane to identify the *cis*- and *trans*- isomers, compare their stability, and draw their chair conformations.

12. Draw the stable conformation of *trans*-1-*tert*-butyl-4-methylcyclohexane, and use the models to explain the reason.

13. Build up molecular models of *cis*- and *trans*- decalin. Decahydronaphthalene can be regarded as the fusion of two cyclohexanes. The dominant conformation of cyclohexane is the chair conformation. There are two ways of fusing the two chair conformations, one is ee fusion and the other is ea fusion. From the models, identify which of these two fusion modes is *cis*- and which is the *trans*- form, then compare the stability of the two isomers.

Questions

1. Please explain the concepts of configuration and conformation, and point out their similarities and differences.

2. Why straight-chain alkanes have the sawtooth shape? Among the isomers of alkanes, the more branched the isomer has, the lower the boiling point it has?

3. Describe the relationship among chiral carbon atoms, enantiomerism, and chiral molecules in organic compound molecules.

2.13 Properties of Organic Compounds

Properties of organic compounds

Chapter 3 The Synthetic Experiments

3.1 Preparation of Cyclohexene

Objectives

1. Study the principle and method of preparing cyclohexene by acid-catalyzed dehydration of cyclohexanol.

2. Master the techniques of fractional distillation and simple distillation with water bath.

Principles

Cyclohexene is usually prepared by cyclohexanol dehydration in an acid catalyzed reaction using strong, concentrated acid, such as sulfuric acid or phosphoric acid. Since sulfuric acid often causes extensive charring in this reaction, phosphoric acid which is comparatively free of this problem, will be used. The reaction is as follows:

$$\text{Cyclohexanol (OH)} \xrightleftharpoons[\triangle]{85\%\ H_3PO_4} \text{Cyclohexene} + H_2O$$

As the reaction is reversible, we use the method of distilling out the binary azeotrope (boiling point 70.8 ℃, water content 10%) formed by the cyclohexene and water out from the system during the reaction in order to enhance its conversion. However, the reactant cyclohexanol can also form a binary azeotrope with water (boiling point 97.8 ℃, water content 80%), so a fractionation device is used to control the distillation temperature at the top of the fractionating column to not exceed 90 ℃.

Apparatus and Reagents

Apparatus: Round-bottom flask, Fractionating column, Heating mantle, Thermometer, Erlenmeyer flask, Beaker, Separatory funnel, Distillation adapter, Condenser, Measuring cylinder.

Reagents: Cyclohexanol, 85% Phosphoric acid, Sodium chloride, 5% Aqueous sodium carbonate solution, Anhydrous calcium chloride.

Procedures

Pour 10.4 mL of cyclohexanol and 4 mL of 85% phosphoric acid into a 50 mL round-bottom flask. Add a magnetic stir bar. Then stir in an electromagnetic stirring heating mantle.

Assemble a fractional distillation apparatus, then use a short fractionating column connected with a condenser and a 50 mL Erlenmeyer flask as the receiving flask. Immerse the receiving flask up to its neck in an ice-water bath to minimize the escape of cyclohexene vapors into the laboratory.

Heat the reaction mixture with a heating mantle until the products begin to boil. The temperature of the distilling vapor must not exceed 90 ℃. The collected liquid in the receiving flask is a mixture of cyclohexene and water. When only a few milliliters of residue remain in the flask, stop the distillation.

Add 1 g solid sodium chloride to the distillate in the receiving flask to saturate the water layer, and shake the flask gently. Decant the distillate into a separatory funnel, add 3-4 mL of 5% aqueous sodium carbonate solution (or 0.5 mL of 20% aqueous sodium hydroxide solution) and shake the separatory funnel gently, then let it stand to stratify. Draw off the lower aqueous layer and then pour the upper layer (crude cyclohexene) through the neck of the separatory funnel into a dry small Erlenmeyer flask. Add about 1-2 g of anhydrous calcium chloride to the flask and swirl the mixture occasionally until the solution becomes clear .

Decant the dry cyclohexene solution into a dry 50 mL of round-bottom flask. Add a stir bar and distill it with a water bath. Use a weighted 50 mL round - bottom flask submerged up to its neck in an ice-water bath as the receiver. Collect the fraction that boils between 80-85 ℃ in the weighted flask, then calculate the percentage yield.

Pure cyclohexene is a colorless liquid, b.p. 82.98 ℃, $n_D^{20}=1.446\ 5$, $d_4^{20}=0.808$.

Notes

1. The cyclohexene (b.p. 83 ℃) will co-distill with the water that is formed (b.p. 70.8 ℃), and the starting material, cyclohexanol, also co-distills with the water (b.p.

97.8 ℃), so the distillation must be done carefully, not allowing the temperature to rise above 90 ℃.

2. The salt minimizes the solubility of the organic product in the aqueous layer.

3. As a small amount of phosphoric acid co-distills with the products, it must be removed by washing with aqueous sodium carbonate solution.

4. Make sure that all the glassware for distilling is clean and dry.

Suggested Waste Disposal

The product is poured into a special recycling bottle. The aqueous solutions (pot residues and washes) should be diluted with water and neutralized before flushing down the drain with excess water. The calcium chloride can be placed in the nonhazardous solid waste container.

Questions

1. Why must the temperature on the top of the fractionating column be controlled during the distillation?

2. What is the purpose of adding NaCl salt before the layers are neutralized and separated?

3. Why must any phosphoric acid be neutralized with sodium carbonate before the final distillation of cyclohexene?

3.2 Preparation of *n*-Butyl Bromide

Objectives

1. Study the principle and preparation of *n*-butyl bromide from *n*-butyl alcohol by treatment with sodium bromide and concentrated sulfuric acid.

2. Learn the technique of reflux with a gas trap apparatus and washing.

Principles

n-Butyl bromide can be prepared by allowing *n*-butyl alcohol to react with sodium bromide and concentrated sulfuric acid.

Main reactions:

$$NaBr + H_2SO_4 \longrightarrow HBr + NaHSO_4$$

$$n\text{-}C_4H_9OH + HBr \xrightleftharpoons{\triangle} n\text{-}C_4H_9Br + H_2O$$

Side reactions:

$$CH_3CH_2CH_2CH_2OH \xrightarrow[\triangle]{H_2SO_4} CH_3CH_2CH{=}CH_2 + CH_3CH{=}CHCH_3$$

$$2CH_3CH_2CH_2CH_2OH \xrightarrow[\triangle]{H_2SO_4} CH_3CH_2CH_2CH_2OCH_2CH_2CH_2CH_3 + H_2O$$

$$2HBr + H_2SO_4 \xrightarrow{\triangle} Br_2 + SO_2\uparrow + 2H_2O$$

Apparatus and Reagents

Apparatus: Round-bottom flask, Condenser, Heating mantle, Thermometer, Erlenmeyer flask, Beaker, Separatory funnel, Distillation adapter, Measuring cylinder, Funnel.

Reagents: *n*-Butyl alcohol, Sodium bromide, Concentrated sulfuric acid, 10% Aqueous sodium hydroxide, Anhydrous calcium chloride.

Procedures

Add 10 mL of water in a 100 mL round-bottom flask, and carefully add 15 mL of concentrated sulfuric acid, mix well, then cool down the mixture to room temperature. 7.4 g (9.2 mL, 0.10 mol) of *n*-butanol and 12.3 g (0.12 mol) of finely ground sodium bromide powder were sequentially added. After shaking sufficiently, add a magnetic stir bar. With the help of a rubber stopper, connect the condenser to a gas trap made with an inverted funnel in the beaker, set up a reflux apparatus, then place some water and sodium hydroxide pellets in the beaker [Ref. Figures 1.4.3 (b) and 1.4.6 (a)].

Heat the mixture with a heating mantle until the mixture begins to reflux gently. Reflux the solution for 0.5 h. Turn off the heating source and cool the mixture down. Remove the condenser with the gas trap.

Assemble a simple distillation apparatus. Distill the mixture until the distillate appears to be clear.

Transfer the distillate to a separatory funnel, add 15 mL of water to it and wash the mixture. Allow the layers to separate. Remove the organic layer and save it in another dry separatory funnel. Wash the organic layer with 6 mL of cold concentrated sulfuric acid and try to separate the lower sulfuric acid layer. Wash the organic layer with 15 mL of water again, then wash with 15 mL of 10% NaOH(to remove traces of acid) and finally wash with 15 mL of water (to remove salts and sodium hydroxide). Check the pH of the water layer with pH paper. It should be close to 7; if not, repeat washing with water.

Dry the cruden-butyl bromide using 1-2 g anhydrous calcium chloride in a small Erlenmeyer flask until the liquid is clear. Decant the clear liquid into a dry 50 mL distilling flask. Distill and collect the fraction that boils between 99-103 ℃. Weigh the product and calculate the percentage yield.

Pure n-butyl bromide is a colorless liquid (b.p. 101.6 ℃, $n_D^{20}=1.440\ 1$, $d_4^{20}=1.275\ 8$).

Notes

1. In the process of reflux, especially when reflux is stopped, pay close attention not to completely bury the funnel in the water to prevent back suction.

2. Check the miscibility in water by adding one drop of distillate to 0.5 mL of water in a test tube.

3. Sulfuric acid is used to get rid of unreacted n-butyl alcohol and the by-products 1-butene, 2-butene and dibutyl ether from crude n-butyl bromide.

4. Be careful to save the organic layer.

Special Instructions

Take special care with concentrated sulfuric acid: It causes severe burns.

Suggested Waste Disposal

The residue from the distillation of 1-bromobutane should be poured into the waste container for halogenated organic solvents. Carefully dilute all non-organic material with water (the reaction pot residue, the sulfuric acid and the sodium hydroxide) and combine and neutralize with sodium carbonate before flushing down the drain with excess water. Place the wet drying agent in a waste container designated for this purpose.

Questions

1. Please explain the role of concentrated sulfuric acid in this experiment.

2. Dibutyl ether, 1-butene and 2-butene can be obtained as by-products in the reactions, explain how to remove them by sulfuric acid in the form of equations.

3. Why must the crude n-butyl bromide be dried carefully with anhydrous calcium chloride prior to final distillation? Why must it be removed before distillation?

3.3 Preparation of Nitrobenzene

Objectives

1. Deepen the understanding of the electrophilic substitution reaction of aromatics through the preparation of nitrobenzene.

2. Master the experimental operations of liquid drying, vacuum distillation and mechanical stirring.

Principles

The nitration reaction is the main method for preparing aromatic nitro compounds, and it is also one of the important electrophilic substitution reactions. The nitration of aromatic hydrocarbons is easier to proceed, usually in the presence of concentrated sulfuric acid with concentrated nitric acid. The hydrogen atoms of the hydrocarbons are replaced by nitro groups to generate the corresponding nitro compounds. The role of sulfuric acid is to provide a strong acidic medium, which is conducive to the generation of nitroxyl cations (N^+O_2). It is a true electrophile. The nitration reaction is usually carried out at a lower temperature. At a higher temperature, the oxidation of nitric acid often leads to the loss of raw materials.

$$C_6H_6 + HNO_3\text{(concentrated)} \xrightarrow[\text{50-55 ℃}]{H_2SO_4\text{(concentrated)}} C_6H_5NO_2 + H_2O$$

Apparatus and Reagents

Apparatus: Reflux condenser, Three-necked round bottom flask, Constant pressure dropping funnel, Mechanical stirrer, Y-shaped tube, Thermometer, Separatory funnel, Vacuum distillation device, Oil bath heating.

Reagents: Benzene, Concentrated nitric acid, Concentrated sulfuric acid, Sodium hydroxide, Anhydrous calcium chloride.

Procedures

1. In a 100 mL Erlenmeyer flask, add 18 mL of concentrated nitric acid. Then slowly add 20 mL of concentrated sulfuric acid under cooling and shaking to make a mixed acid for later use.

2. Install the device as shown in Figure 3.3.1.

3. Put the prepared mixed acid in a constant pressure dropping funnel, and put 18 mL of benzene in a 250 mL three-necked round bottom flask.

4. Start stirring, slowly drip the mixed acid into the reaction flask, and control the reaction temperature at 50-55 ℃.

5. Cool the reaction mixture in a cold water bath, then transfer it to a 100 mL separatory funnel, release the lower layer (mixed acid). The organic layer is washed with an equal volume (about 20 mL) of water, 5% sodium hydroxide solution and water. The nitrobenzene was transferred into a 50 mL conical flask containing 2 g of anhydrous calcium chloride, and vortexed until the turbidity disappeared.

6. Filter the dried nitrobenzene into a 50 mL dry round-bottomed flask, connect to an air condenser, heat and distill on an asbestos net, collect the 205-210 ℃ fraction to obtain the product nitrobenzene.

Special Instructions

1. Nitro compounds are more toxic to the human body, so be careful when handling

nitro compounds. If you accidentally touch the skin, wash it with a small amount of ethanol immediately, and then wash with soap and warm water.

2. When washing nitrobenzene, especially NaOH should not be vigorously shaken, otherwise the emulsification of the product will be difficult to stratify. In this case, solid NaOH or NaCl saturated solution can be added to a few drops of alcohol to stand for a while to stratify.

3. Since the nitrobenzene remaining in the flask is prone to violent decomposition at high temperature, do not distill the product to dryness or make the temperature exceed 114 ℃.

4. The nitrification reaction is an exothermic reaction, and the temperature should not exceed 55 ℃.

Suggested Waste Disposal

Residues of flammable and explosive substances (such as metallic sodium, white phosphorus, match heads) shall not be poured into a dirt bucket or a sink, and should be collected in designated containers. Waste liquid, especially strong acids and alkalis, cannot be poured directly into the sink. It should be diluted first, then poured into the sink, and then rinse the sink and sewer with a large amount of tap water. Poisons should be collected after going through the approval procedures in accordance with the regulations of the laboratory, strictly operated during use, and handled properly after use.

Questions

1. Why should the reaction temperature be controlled between 50-55 ℃ in this experiment? What are the disadvantages if the temperature is too high or too low?

2. What is the purpose of washing the crude product with water, lye and water in sequence?

3.4 Preparation of *p*-Nitrobenzoic Acid

Objectives

1. Master the principle and method of preparing *p*-nitrobenzoic acid from *p*-nitrotoluene.

2. Practice and learn the purification method of solid acid products.

Principles

The side-chain oxidation method is commonly used to prepare aromatic acids. Potassium dichromate (sodium) or potassium hyperbolic acid; sulfuric acid are used to oxidize alkyl groups on the benzene ring to carboxyl group. The resulting crude product is an acidic solid substance which can be purified by adding alkali to dissolve and then acidifying. The reaction is as follows:

$$O_2N-C_6H_4-CH_3 + Na_2Cr_2O_7 + 4H_2SO_4 \longrightarrow O_2N-C_6H_4-COOH + Cr_2(SO_4)_3 + Na_2SO_4 + 5H_2O$$

Apparatus and Reagents

Apparatus: Round bottom flask, Reflux condensing tube, Electromagnetic heating stirrer, Dropping funnel, Suction flask, Beaker.

Reagents: *p*-Nitrotoluene, $Na_2Cr_2O_4$, Concentrated sulfuric acid, 15% Sulfuric acid, 5% NaOH solution.

Procedures

1. Install the device with stirring, reflux and dripping to a 100 mL three-necked flask, add 3 g of *p*-nitrotoluene, 9 g of potassium bicarbonate powder and 20 mL of water in sequence.

2. Drop 12.5 mL of concentrated sulfuric acid from the dropping funnel with stirring. (Pay attention to cooling with cold water. Avoid condensation of *p*-nitrotoluene on the condenser due to high temperature volatilization.)

3. After the sulfuric acid is dripped, heat to reflux, and the reaction solution turns black. (During this process, white *p*-nitrotoluene may be precipitated in the condenser tube, so the condensed water can be appropriately turned off so that it melts and drips.)

4. After the reactant is cooled, add 40 mL of ice water with stirring, and there will be precipitation, filter under reduced pressure and wash with 25 mL of water twice.

5. Put the washed black solid of *p*-nitrobenzoic acid into 15 mL of 5% sulfuric acid, heat it on a boiling water bath for 10 min, and then filter with suction after cooling (The purpose is to remove the unreacted salt).

6. Dissolve the solid after suction filtration in 25 mL of 5% NaOH solution, warm it to 50 ℃ and then suction filter. Add 0.5 g of activated carbon to the filtrate, boil and then suction filter while hot. (This step is very critical, if the temperature is too high, the *p*-nitrotoluene will melt and be filtered into the filtrate. If the temperature is too low, the sodium *p*-nitrobenzoate will precipitate, which will affect the purity or yield of the product.)

7. With sufficient stirring, slowly add the filtrate obtained by suction filtration into a beaker containing 30 mL of 15% sulfuric acid solution to precipitate yellow precipitate, suction filtration, wash twice with a small amount of cold water, and weigh

after drying. (The order of addition cannot be reversed, otherwise the product will be impure.)

8. Recrystallization with alcohol-water mixed solvents. The melting point of *p*-nitrobenzoic acid is 241-242 ℃.

Special Instructions

1. Start from the dropping of concentrated sulfuric acid, and keep stirring during the whole reaction process.

2. When adding concentrated sulfuric acid dropwise, only stir without heating; the speed of adding concentrated sulfuric acid should not be too fast, otherwise it will cause a violent reaction.

3. After being transferred to 40 mL cold water, a small amount (about 10 mL) of cold water can be used to wash the flask again.

4. When using alkali to dissolve, it can be warmed appropriately, but the temperature should not exceed 50 ℃, In case the unreacted p-nitrotoluene melts and enters the solution.

5. When acidifying, pour the filtrate into the acid, and do not pour the acid into the filtrate the other way around.

6. The purified product is dried in a steam bath.

Suggested Waste Disposal

Treatment of acid and alkali waste liquid. The amount of various acid-base solutions in the laboratory is relatively large. After use, they should be collected according to their acid-base properties and treated by neutralization. In the case where it is found that the two waste liquids can be mixed with each other, they can be neutralized with each other in small amounts, and then diluted with water to reach the standard and discharged.

Treatment of waste liquid containing oxidizing agent and reducing agent. The treatment of this kind of waste liquid often adopts the oxidation-reduction method. First, the waste liquid containing the oxidizing agent and the reducing agent is collected separately. Then carry out harmless treatment.

Questions

1. What are the oxidation methods for aryl side chains? What are the laws of oxidation?

2. In addition to electric stirring, what measures are there to improve heterogeneous reaction?

3. Why do we need to pour the filtrate into the acid during acidification, but not the other way around?

3.5 Green Synthesis of Adipic Acid

Objectives

1. Through the preparation of adipic acid, understand the disadvantages of traditional synthetic methods and the advantages of green synthetic methods.

2. Be familiar with the recycling use of catalysts without recycling.

Principles

Adipic acid is commonly known as fatty acid, and its molecular formula is $C_6H_{10}O_4$. For the production process of adipic acid, the most widely used in the world is the nitric acid oxidation process route using cyclohexanol or cyclohexanone as raw materials. In this production route, strong oxidizing nitric acid is used, which severely corrodes equipment, and the N_2O gas produced during the production process is considered to be one of the causes of global warming and ozone reduction, causing great pollution to the environment. Japanese scientist Ryoji Noyon published an article on the green synthesis of adipic acid in Science in 1998. He proposed a green preparation route using water as solvent, H_2O_2 as oxidant, sodium tungstate (Na_2WO_4) as catalyst. And with the participation of potassium hydrogen sulfate ($KHSO_4$), methyl trioctylammonium chloride (aliquat 336) is used as the phase transfer catalyst. This route does not use strong acid, does not produce N_2O harmful gas, uses water as a solvent and does not produce waste liquid, and the catalyst can be recycled directly without recycling. Since then, some researchers have discovered that no phase transfer catalyst is needed, water is used as solvent, H_2O_2 is used as catalyst, and adipic acid can also be synthesized using a coordination catalyst of sodium tungstate-oxalic acid in situ synthesis. The reaction is as follows:

$$\text{cyclohexene} \xrightarrow[\text{KHSO}_4\text{, Aliquat 336}]{\text{NaWO}_4\cdot 2\text{H}_2\text{O}\cdot \text{H}_2\text{O}_2} \text{HOOC(CH}_2)_4\text{COOH}$$

In this experiment, cyclohexene is used as raw material, and under the catalysis of sodium tungstate-oxalic acid, adipic acid can be synthesized by hydrogen peroxide oxidation. The reaction is as follows:

$$\text{cyclohexene} \xrightarrow[H_2C_2O_4 \cdot 2H_2O \cdot H_2O_2]{NaWO_4 \cdot 2H_2O} HOOC\text{-}CH_2CH_2CH_2CH_2\text{-}COOH$$

Apparatus and Reagents

Apparatus: Magnetic heating stirrer, Condenser, Round bottom flask, Beaker, Drying tube, Watch glass, Alkali burette, Melting point tester, Three-neck flask, Vacuum pump, Analytical balance.

Reagents: Sodium tungstate (AR), Cyclohexanol (AR), Cyclohexene (AR), Concentrated nitric acid (AR), Potassium hydrogen sulfate (AR), Methyl trioctyl chloride (AR), Hydrogen peroxide (AR), Oxalic acid (AR).

Procedures

1. Traditional preparation method

Add 2 mL of concentrated nitric acid into a 10 mL round bottom flask, put in a stirring magnet, and install a condenser. The device diagram is as follows (see Figure 3.5.1). Heat to 80 ℃ in a fume hood (for N_2O gas release). Then add 1 mL of cyclohexanol slowly from the top of the condenser tube to the round bottom flask through a dropper. In the process of dropping, try to avoid contact between cyclohexanol and the inner wall of the condenser tube, and the dropping rate should be controlled at about 1 drop/min. The dripping process lasts for 30-40 min. After the dropwise addition, keep the temperature at 80 ℃ and continue the reaction for 2 h, then cool to room temperature, and cool in an ice-water bath to precipitate crystals. Filter with a Büchner funnel, wash with a small amount of ice water, and then with a small amount of ethanol to drain the crystals as much as possible. The products were collected and dried in a drying oven for 15 min, weighed to calculate the yield, and a small amount of the product was taken to determine the melting point.

2. Green preparation method Sodium tungstate-oxalic acid is used as catalyst

In a 100 mL three-necked flask, 1.50 g sodium tungstate, 0.57 g oxalic acid, 34.0 mL 30% H_2O_2 and magneton were added. After stirring at room temperature for 15-20 min, 6.00 g of cyclohexene was added. Set up the reflux device (see Figure 3.5.2), continue to stir quickly and vigorously and heat to reflux. During the reaction, the reflux temperature will slowly increase. After refluxing for 1.5 h, pour the hot reaction solution into a 100 mL beaker, cool the cold water to near room temperature, and then cool it in an ice-water bath for 20 min. Filter with a Büchner funnel, wash with 5 mL of ice water, and then wash with 5 mL of ethanol. Drain the crystals as much as possible. After drying the product to equilibrium, weigh it, record the mass of adipic acid, calculate the yield, and take a small amount of product for determination of melting point.

3. Determination of adipic acid content

The content of adipic acid was determined by acid-base titration. Weigh 0.1 g (accurately to 0.0001g) of the experimental product twice into a 250 mL Erlenmeyer flask, add 50 mL of hot distilled water, and stir to dissolve the sample. Add 1 drop of phenolphthalein indicator and titrate with 0.1 mol/L sodium hydroxide standard solution to reddish color. The end point is that it does not fade within 30 s. Titrate 2 parts each. Calculate the percentage of adipic acid in the sample.

Results

The results are recorded in Table 3.5.1.

Table 3.5.1 Data record

Preparation	Quality of raw materials/g	Theoretical output/g	Actual output/g	Yield/%	Melting point/℃	Purity /%
Traditional method						
Green method						

Special Instructions

1. This reaction is exothermic. After the reaction starts, the mixture will exceed 45 ℃. If the temperature of the mixture cannot rise to 45 ℃ after starting the reaction at room temperature for 5 min, it can be carefully heated to 40 ℃ to start the reaction.

2. Keep shaking or stirring, otherwise it will easily burst out of the container.

3. In order to increase the yield, it is best to cool the solution with ice water to reduce the solubility of adipic acid in water.

Suggested Waste Disposal

The treatment of waste liquid containing methanol, ethanol and acetic acid can be recycled after being refined by distillation. If the concentration is low, it can be diluted with a large amount of water and discharged after reaching the standard.

Question

In this experiment, in the green preparation, because the raw materials were added too early, the reaction proceeded in advance, which eventually resulted in a lower yield and a longer melting range.

3.6 Preparation of Cinnamic Acid

Objectives

1. Learn the principle and method of preparation of cinnamic acid.
2. Master the operations of reflux and steam distillation.

Principles

When benzaldehyde is mixed with an anhydride and heated under the corresponding carboxylate catalyst, an α, β-unsaturated acid can be prepared through the Perkin reaction. The main role of the catalyst is to promote enolization of the anhydride.

In this experiment, potassium carbonate are used to substitute for potassium acetate in order to shorten the reaction time and improve the product yield. The reaction is as follows:

$$C_6H_5CHO + (CH_3CO)_2O \xrightarrow[140\text{-}180\ ^\circ C]{K_2CO_3} C_6H_5CH{=}CHCOOH + CH_3COOH$$

Apparatus and Reagents

Apparatus: Round-bottom flask, Condenser, Distillation adapter, Heating mantle, Büchner funnel, Beaker, Measuring cylinder, Suction flask, Water vacuum pump, Filter paper.

Reagents: Benzaldehyde, Acetic anhydride, Anhydrous potassium carbonate, Concentrated hydrochloric acid, 10% Sodium hydroxide solution, Congo red test paper.

Procedures

Place 3 mL (0.03 mol) of benzaldehyde, 8 mL (0.085 mol) of acetic anhydride, 4.2 g (0.03 mol) grinded anhydrous potassium carbonate and 2-3 boiling stones into a 250 mL round-bottom flask. Among these materials, benzaldehyde and acetic anhydride are both redistilled before use. Heat the mixture to reflux with a heating mantle for about 30 min. Because carbon dioxide will be released in the process, bubbles will be produced at an early stage of the reaction.

Cool down the reaction mixture, and add 20 mL warm water into the flask. Distill excessive benzaldehyde from the flask by means of steam distillation. Then cool the flask down and add 10 mL 10% sodium hydroxide solution. Make sure all the cinnamic acid is converted into the sodium salt and dissolves completely. Filter the solution into a

250 mL beaker and cool down to room temperature, stir and acidify it with concentrated hydrochloric acid and check it with Congo red test paper until it becomes blue. Filter the solution and wash the solid with an appropriate amount of cold water. Keep the aspirator running until the filter residue becomes dry. Air-dry the crude product.

The crude product can be recrystallized with hot water or the mixture of water and ethanol (the ratio is 5 : 1, *V*/*V*). Weigh the product and calculate the yield.

Pure cinnamic acid(*trans*- form) is white crystal with m.p.133-136 ℃.

Notes

1. Benzaldehyde may be automatically oxidized to benzoic acid if it is stored for a long time. Benzaldehyde should be redistilled before use, otherwise the existence of trace benzoic acid not only influences the progress of reaction, but also is difficult to be removed. The fraction of 170-180 ℃ is collected for use.

2. Acetic anhydride may hydrolyze to acetic acid when it is exposed to moisture for a long time and therefore it should be redistilled before use as well.

3. Cinnamic acid has *cis*- and *trans*-isomers, the product prepared usually is *trans*-form with a melting point of 133-136 ℃.

Suggested Waste Disposal

All filtrates should be poured into a waste container designated for non-halogenated organic waste.

Questions

1. What compound is removed by using steam distillation? Why can steam distillation purify the crude product?

2. What is the product from the reaction of benzaldehyde with propionic anhydride in the presence of anhydrous potassium carbonate?

3. Can sulfuric acid be used for acidification?

3.7 Preparation of Ethyl Acetate

Objectives

1. Comprehend mechanism and protocol of esterification reaction.

2. Learn reflux operation and consolidate distil and extraction technique.

Principles

Esterification reaction refers to the process in which acid and alcohol react to produce ester and water, and the reverse reaction is called ester hydrolysis reaction.

Esterification reactions generally use strong acids as catalysts, such as concentrated sulfuric acid, dry hydrogen chloride, strong organic acids and cation exchange resins. Among them, concentrated sulfuric acid is the most commonly used. The purpose of adding strong acids is to quickly reach equilibrium. The reaction is as follows:

$$R-\overset{\overset{\displaystyle O}{\|}}{C}-OH + R'OH \xrightleftharpoons{H_2SO_4} R-\overset{\overset{\displaystyle O}{\|}}{C}-OR + H_2O$$

In order to increase the yield of the reaction, the alcohol or acid is often used in excess, or the ester and water produced by the reaction are removed, so that the balance shifts to the right.

In this experiment, excess ethanol was used to increase the reaction yield. Ethyl acetate and water form an azeotropic mixture (70.4 ℃), which has a lower boiling point than ethanol (78 ℃) and acetic acid (118 ℃). Ethyl acetate has a boiling point of 77.06 ℃, so it is easy to evaporate.

In addition to the formation of ethyl acetate, it also generates a side reaction of ether. The reaction is as follows:

$$CH_3COOH + CH_3CH_2OH \xrightleftharpoons{H_2SO_4} CH_3COOC_2H_5 + H_2O$$

Apparatus and Reagents

Apparatus: 50, 100 mL Round-bottom flash, 100 mL Separating funnel, Allihn condenser, 50 mL Conical flask, 25 mL Measuring cylinder, Distil devices.

Reagents: Absolute alcohol, Acetic acid, Concentrated sulfuric acid, Saturated NaCl solution, Saturated $CaCl_2$ solution, Saturated Na_2CO_3 solution, Dry Na_2SO_4.

Procedures

Weigh 20 mL absolute alcohol (15.7 g, 0.34 mol), 12 mL acetic acid (12.59 g, 0.21 mol) into a 100 mL round-bottom flask, and slowly add 10 mL concentrated sulfuric acid under shaking. After homogenous, add 2-3 zeolite and set Allihn condenser, followed by reflux in water bath for 30 min. After reaction finishes, add 2-3 more zeolite, change the equipment to the distil mode. Distil the reaction solution until no clear liquid flow out. Until this step, crude product is given.

Add saturated Na_2CO_3 solution into the crude product until the organic layer is neutralized. Transfer solution into a 100mL separation funnel and save the organic layer, which is further washed by equal volume of saturated NaCl solution for twice and saturated $CaCl_2$ twice. Transfer the organic layer into the dry conical flask, which is dried by $MgSO_4$.

Filter the obtained ethyl acetate and subject the liquid to distil device with a 50 mL round-bottom flask and collect the product at temperature of 75-80 ℃.

Pure ethyl acetate is colorless liquid, b.p. 77.06 ℃, density 0.901, slightly lighter than water, microsolubility in water with fruit smell.

Notes

1. When temperature is high than 120 ℃, side products would be increased. Thus in distil experiment, it should be maintained at slight boiling state.

2. Distilled liquid contain ester, water, few alcohol and acetic acid. Using Na_2CO_3 to remove acetic acid and H_2SO_3 (reduced product of sulfuric acid), $CaCl_2$ to remove alcohol. However, the Na_2CO_3 should not be access or gel would be formed in the process of $CaCl_2$ treatment.

Suggested Waste Disposal

Any aqueous solutions should be placed in a container specially designated for dilute aqueous waste. Place any excess ester in the nonhalogenated organic waste container.

Questions

1. What is role of sulfuric acid?

2. What kinds of side products in distilled ethyl acetate?

Further Reading

Flavor and fragrance esters

3.8 Preparation of Acetylsalicylic Acid

Objective

Learn the experimental methodsand principles of aspirin preparation.

Preparation of acetylsalicylic acid (experimental principle)

Preparation of acetylsalicylic acid (operation video)

Principles

Acetylsalicylic acid, commonly called aspirin, is synthesized from salicylic acid (ortho-hydroxybenzoic acid) and acetic anhydride. Until now, aspirin is still widely used as an antipyretic, analgesic and anti-inflammatory drug. Salicylic acid is a bifunctional compound with a phenolic hydroxyl group and a carboxyl group. Two different esterification reactions can be carried out. When reacted with acetic anhydride, acetylsalicylic acid, namely aspirin, can be obtained; if it reacts with excess methanol, methyl salicylate is produced. This experiment will carry out the previous reaction experiment.

The reaction formula:

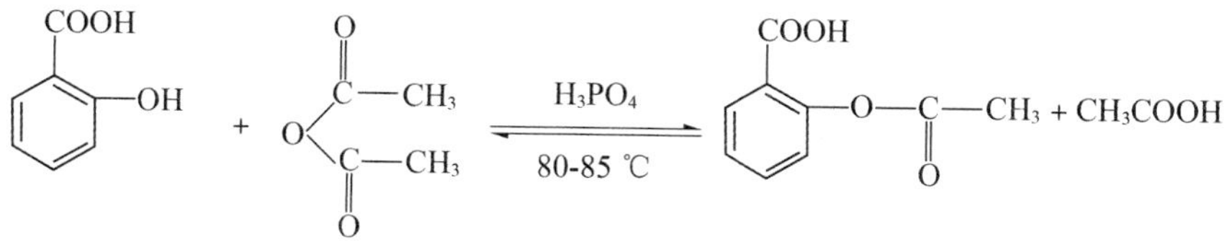

$$\text{COOH, OH} + \text{O(C(=O)—CH}_3)_2 \xrightarrow[80\text{-}85\ ^\circ\text{C}]{H_3PO_4} \text{COOH, —O—C(=O)—CH}_3 + CH_3COOH$$

While generating acetylsalicylic acid, a condensation reaction occurs between salicylic acid molecules to generate a small amount of polymer.

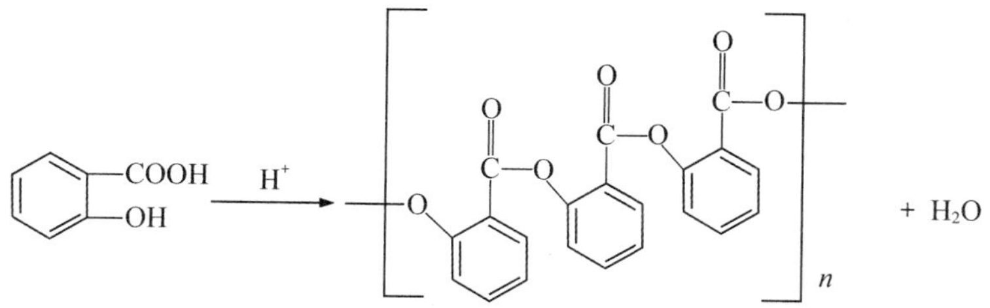

$$\text{—COOH, —OH} \xrightarrow{H^+} [\text{—O—C(=O)—O—C(=O)—O—C(=O)—O—}]_n + H_2O$$

Acetylsalicylic acid can react with sodium bicarbonate to form a water-soluble sodium salt, while the by-product polymer cannot be dissolved in sodium bicarbonate. This difference in properties can be used for the purification of aspirin.

The impurity that may be present in the final product is salicylic acid itself, which is caused by the incomplete acetylation reaction or the hydrolysis of the product during the separation step. It can be removed in each step of purification process and product recrystallization. Like most phenolic compounds, salicylic acid can form a dark complex with ferric chloride. Because aspirin has been acylated with the phenolic hydroxyl group, it no longer reacts with ferric chloride in color, so impurities are easily detected out.

Apparatus and Reagents

Apparatus: SHZ-C type circulating water multi-purpose vacuum pump (public), Electric heating constant temperature water bath, Drying oven, Shelf pan medicine balance, Stainless steel spatula, Scissors, 50 mL Conical flask, Thermometer (0-100 ℃), 10 mL Measuring cup, 50 mL Beaker, Suction filter flask, Büchner funnel, Filter paper, Weighing paper, Medicine spoon, Test tube rack and test tube, Recycling bottle, Wooden clamp, Glass rod, Ice, Spherical condenser, 50 ml round bottom flask.

Reagents: Salicylic acid, Acetic anhydride, Phosphoric acid, 95% ethanol, 0.1% Ferric chloride solution, Concentrated hydrochloric acid, Saturated sodium bicarbonate solution.

Procedures

Add 3 g (0.022 mol) of dry salicylic acid to a dry 50 mL flask, then slowly add 4.5 g (0.04 mol) of acetic anhydride, shake well, and add 5 drops of phosphoric acid dropwise. Shake well to dissolve all salicylic acid. Put the flask and the condenser into a water bath at 80-85 ℃ and keep it at a constant temperature for 10-15 min with constant shaking during the period. Cool slightly, pour the reactant into 50 mL water under constant stirring, cool with cold water, filter with suction, and wash with appropriate amount of cold water. The crude product after suction filtration was transferred to a 100 mL beaker, and 38 mL of saturated sodium bicarbonate aqueous solution was added with stirring. Continue to stir for a few minutes until no carbon dioxide is produced. Suction filtration to filter out the by-product polymer, and rinse the funnel with 5-10 mL of water, pour the filtrate into a beaker containing 7 mL of concentrated hydrochloric acid and 15 ml of water in advance, stir evenly. That is, acetylsalicylic acid crystals will precipitate out, put the beaker Cool with ice-cold water to make the crystals crystallize completely. Filter with suction and wash the crystals with cold water. Transfer the crystals to a watch glass, and weigh after drying, calculate yield. Take a few crystals and dissolve them in a test tube containing 1 mL of 95% ethanol, add 1-2 drops of 0.1% ferric chloride solution, observe for color changes, and determine whether there is unreacted salicylic acid in the product.

Notes

1. Acetic anhydride should be freshly steamed.

2. Acetylsalicylic acid is easily decomposed by heat, so its melting point is not obvious. Its decomposition point is 128-135 ℃. When determining the melting point, the carrier should be heated to 120 ℃ first, and then the sample should be added for determination.

Special Instructions

1. The instrument should be completely dried, the medicine should be dried, and the acetyl anhydride should be freshly steamed, otherwise the yield will be very low.

2. The heating temperature of the water bath should not be too high and the time should not be too long, otherwise the by-products may increase.

3. If it does not crystallize, use a glass rod to rub the bottle wall or place it in ice water to cool.

4. The heating time for recrystallization should not be too long, the water temperature should be controlled, and the product should be dried naturally.

Suggested Waste Disposal

After the experiment, the product is recycled and the instrument is cleaned, the waste liquid is recycled into the waste liquid bucket.

Questions

1. Why phosphoric acid is needed in the acetylation reaction?

2. What side reaction will occur in this experiment?

3. Could you list some ways to test the purity of aspirin?

Further Reading

Analgesics

3.9 Preparation of Acetanilide

Objectives

1. Learn the principle and method of preparing acetanilide by glacial acetic acid as the acetylating reagent.

2. Master the experimental technique of recrystallization.

Principles

Acetylation is often used to "protect" a primary or a secondary amine functional group. The amine group can be regenerated readily by hydrolysis in acid or base.

Acetanilide can be prepared from aniline in several ways, using acetyl chloride, acetic anhydride or glacial acetic acid as starting materials. Acetanilide is not only the first analgesic-antipyretic drug (relieving pain and fever) with another name of "antipyretic ice", but also an important itermediate in the synthesis of sulfa drugs. Acetyl chloride reacts very vigorously. Acetic anhydride is preferred for a laboratory synthesis because its rate of hydrolysis is low enough to allow the acetylation of amine to be carried out in aqueous solution. It gives a product of high purity and in good yield. The procedure with glacial acetic acid is not only economical, but also more environment-friendly. In the present experiment, we use glacial acetic acid as acylation agent to prepare the acetanilide.

$$C_6H_5NH_2 + CH_3COOH \xrightarrow{Zn} C_6H_5NHCOCH_3$$

Apparatus and Reagents

Apparatus: Round-bottom flask, Fractionating column, Distillation adapter, Condenser, Heating mantle, Thermometer (150 ℃), Suction flask, Büchner funnel, Measuring cylinder, Filter paper,Beaker, Water vacuum pump.

Reagents:Freshly distilled aniline, Glacial acetic acid, Zinc powder.

Procedures

Place 10 mL (10.2 g, 0.11 mol) of aniline into a 100 mL round-bottom flask, and add 15 mL (15.7 g, 0. 26 mol) of glacial acetic acid and 0.1 g of zinc powder to the flask. Assemble a fractionating column with a thermometer, and use a 50 mL round-bottom flask as a collector, as shown in Figure 3.9.1.

Heat the round-bottom flask slowly to reflux. The reaction mixture remains slightly boiling for about 15 min. Then control the temperature of the distilling vapor being about 105 ℃, while water produced by the reaction and a small amount of glacial acetic acid can be distilled out. After 1.5 h, the reaction finished. With stiring,the hot solution is poured into a beaker with 200 mL of ice-cold water to complete the crystallization process. Collected the crystals by suction filtrate, wash with ice-cold water, recrystallize from water, calculate yield and determine the melting point of acetanilide.

Pure acetanilide is a colorless sheet shaped crystal, m.p. 113-114 ℃.

Notes

1. Aniline is a toxic substance, and it can be absorbed through the skin. Take care in handling it. Aniline becomes dark after a long-term storage because of being oxidized, so it's better to use redistilled aniline.

2. The role of zinc powder is to prevent the oxidation of aniline. Only a small amount of zinc is needed, if too much is added, water-insoluble zinc hydroxide will be

formed.

3. Do not allow the fractionation temperature rising too high, otherwise a large amount of acetic acid will be distilled out.

Suggested Waste Disposal

Aqueous solutions obtained from filtration operations or crystallization steps should be poured into the container designated for aqueous wastes.

Questions

1. Why should the temperature of the upper end of the fractionating column controlled at about 105 ℃ during reaction?

2. Why use the ice-cold water rather than room temperature water when washing the final product?

3. Why is acylation usually carried out first when aniline is used as the reactant for some substitution reactions on benzene rings?

Chapter 4 Comprehensive and Designed Experiments

4.1 Isolation of Caffeine from Tea Leaves

Objectives

1. Know the general method of alkaloidextraction from plant drug through the isolation of caffeine from tea.

2. Learn the operation of Soxhlet extractor and sublimation technique for purifying organic compounds.

3. Learn the general identification methods of alkaloids.

Isolation of caffeine from tea leaves(experimental principle)

Isolation of caffeine from tea leaves(operation video)

Principles

Caffeine is an alkaloid, a class of naturally occurring compounds containing nitrogen and having the properties of an organic amine base. Tea and coffee are not the only plant sources of caffeine. Others include kola nuts, guarana seeds, and in small amount, cocoa beans. Caffeine belongs to family of naturally occurring compounds called xanthines. In this experiment, caffeine will be extracted from tea leaves. The major problem of the extraction is that caffeine does not occur alone in tea leaves, but is accompanied by other natural substances from which it must be separated.

There are many derivatives of xanthine (2,6-dihydroxypurine) and a large amount of tannin in the tea leaves, such as caffeine, theophylline and theobromine. The amount of caffeine in tea varies from 2% to 5%. Caffeine is an odorless, bitterness and white crystal, melting point is 238 ℃. Caffeine is water soluble and is more soluble in hot water, chloroform. The methods for isolation of caffeine from tea mainly include continuous extraction and leaching.

In the continuous extraction method, Soxhlet extractor is employed with ethanol as the solvent, the solvent is eventually evaporated to obtain crude caffeine. Crude caffeine also contains some other alkaloids and impurities (such as tannins, etc.), which can be further purified through sublimation. The product after sublimation usually obtains higher purity at the cost of loss of some product. Caffeine containing crystal water loses crystal water when heated to 100 ℃ and begins to sublimate. It sublimates significantly at 120 ℃, and rapidly at 176 ℃.

In addition, caffeine can be extracted from the hot water solution of tea. Because tannins are water soluble, so a large amount of tannins are extracted together. Then add the $Pb(AC)_2$ to precipitate the tannin from the solution of the tea.

Then caffeine can be extracted from the tea solution with ethyl acetate, but other elements are not ethyl acetate soluble and remain behind in the aqueous solution.

The extracted steps of caffeine are as Figure 4.1.1.

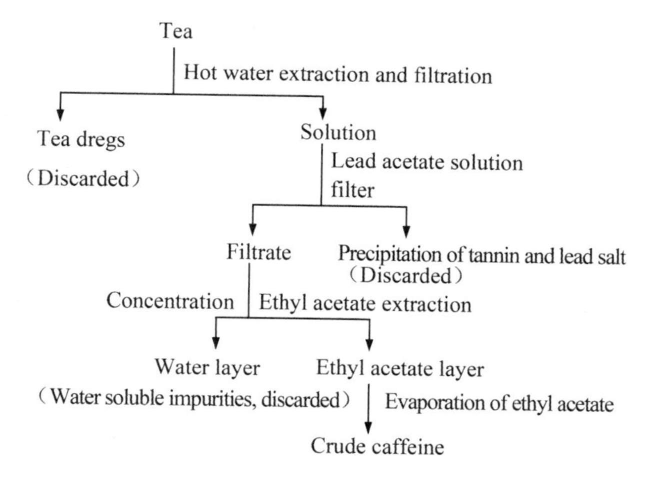

Figure 4.1.1 A flowchart for the isolation of caffeine

Apparatus and Reagents

Apparatus: Soxhlet extractor, Distillation device, Filtration apparatus, Thermostatic bath, Electromagnetic furnace, Separatory funnel, Funnel, Evaporating dish, Crucible, Beaker, Cylinder, Tube, Filter paper, Cotton, Boiling stone.

Reagents: Tea(commercially available), 10% Lead acetate solution, Ethyl acetate, Saturated NaCl solution, Potassium chlorate, Concentrated hydrochloric acid, Concentrated ammonia water, Potassium bismuth iodide reagent, 5% Sulfuric acid solution, 95% Ethanol, Quicklime (CaO).

Procedures

1. Extraction of caffeine

Method 1:Continuous extraction

(1) Weigh 8 g of tea and put them into the filter paper cylinder of the Soxhlet extractor, add 80 mL of 95% ethanol and a few boiling stones into the flask. Install the Soxhlet extractor (Figure 4.1.2), turn on the condensed water, heat to reflux, and continue to extract for 1-1.5 h (when the color of the extract turns very light, stop the extraction), when the condensate is just siphoned down, stop heating immediately and cool it down.

(2) Install distillation device, heat and recover most of the ethanol. Pour the residual liquid (approximately 15-20 mL) into the evaporating dish, wash the distillation flask with a small amount of 95% alcohol, combine the washing liquid in the evaporating dish, and concentrate on the heating mantle to around 10 mL of the residual liquid.

(3) Add 4 g quicklime (CaO) powder into the concentrated residual liquid, stir evenly, and evaporate to dryness on the heating mantle. First fix the evaporating dish on the iron ring, then heat it and evaporate it into a loose shape under continuous stirring (crush the blocks, bake over low heating power and remove all the water), so as to remove all the water. After cooling, wipe off the powder stained on the edge of the evaporating dish to avoid polluting the product during sublimation.

(4) Cover the above-mentioned evaporating dish with a round filter paper pierced with many small holes, and cover a dry glass funnel on the top (a little absorbent cotton is put on the neck of the funnel to reduce the escape of caffeine vapor) (Figure 4.1.3). Use the electric heating mantle to heat the evaporating dish for sublimation. When white needle crystals appear on the filter paper, stop heating. After cooling (about 5 min), carefully uncover the funnel and filter paper, and carefully scrape the caffeine crystals (white, needle-like) attached to the filter paper and around the vessel into a dry, clean, weighed 50 mL beaker with a knife. After the residue is mixed, use a larger heating power to continue heating and sublimation once (or twice). Combine each sublimation to collect caffeine crystals and weigh them.

Method 2: Leaching extraction

(1) Place 5 g of dry tea leaves, and 100 mL of hot water in a 200 mL beaker. The liquid is heated to boil for about 15 min. The filter cotton is used to filtrate the solution and the dreg of tea is discarded. Add 15-20 mL of 10% $Pb(AC)_2$ solution under stir into the hot filtrated solution until the precipitate is not formed again.

(2)Heating the above solution about 5 min, then the solution is filtrated by suction filtration. Cool the solution, suction filtration again if the precipitate is formed proceed.

(3)Place the above solution in a separatory funnel. Add 25 mL of ethyl acetate and 15 mL of saturated NaCl aqueous solution into the separatory funnel, then strongly shake it. When shaking the separatory funnel, the ethyl acetate vapor is formed, because the ethyl acetate is easily volatile. To release this vapor, the funnel is vented by holding it upside-down and slowly opening the stopcock. Then the funnel is placed for a moment and the top stopper is immediately removed.

The water and ethyl acetate will separate into two layers after a short time. And the ethyl acetate (the upper layer) may be separated, pour it into a small distilling flask. Add 2-3 boiling stones.

Assemble an apparatus for simple distillation and remove the ethyl acetate by distillation, using a water bath to heat for vaporization of ethyl acetate. The remainder is the crude caffeine.

2. Identification of caffeine

(1)Place some crystals of crude caffeine in a crucible and add small amount of $KClO_3$ crystals and 2-3 drops of concentrated HCl.

To heat on the asbestos wire gauze until the liquid is completely vaporized. Cool and add 1 drop of concentrated NH_4OH, and then the violet color appears. This test is used for identifying the presence of alkaloid. This reaction is referred to as ammonium violurate reaction.

(2)Place 2 mL of 5% H_2SO_4 into remainder of the crude caffeine. To solve with stir, take about 1 mL and add 2 drops of Dragendoll agent. If the orange precipitate is formed, this result means that the alkaloid is present.

Notes

1. Sublimation is a process in which a solid substance with a higher vapor pressure is heated below its melting point and then directly turn into vapor without passing through a molten state. The vapor directly turns to a solid when getting cold. Sublimation is one of the methods to purify solid organic compounds. Substances that can be purified by sublimation must meet the following conditions: ①The solid have a higher vapor pressure; ② There should be a significant difference between the vapor pressure of the impurity and the organic compound to be purified.

2. Ammonium violurate reaction is shown as follows.

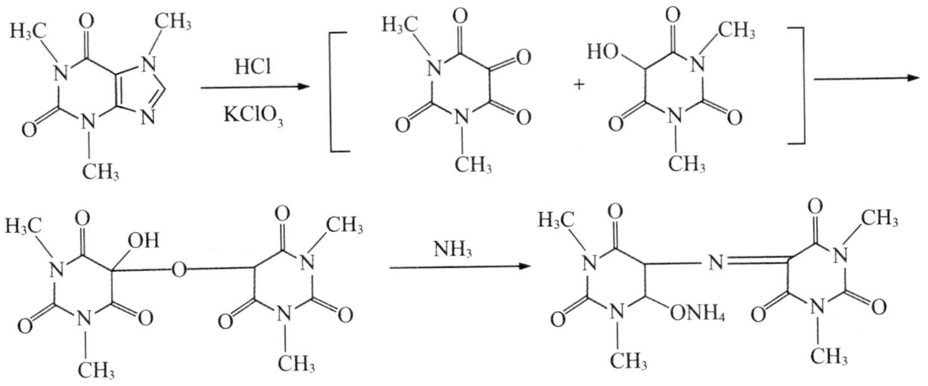

(violet color)

Tetramethyl ammonium violurate

3. The filter paper should be close to the wall of the device and be easy to remove. The height of the extract cannot exceed the siphon, otherwise the extract cannot be fully soaked by the solvent, which would affect the extraction effect. The tea should not leak out of the filter paper tube, so as not to block the siphon tube.

4. Quicklime (CaO) powder is used for absorption of water and neutralization in order to remove impurities.

5. Ethyl acetate is a colorless volatile liquid, with a strong ether-like odor, slightly fruity, easy diffusion, not lasting, soluble in water, miscible in any proportion with petroleum ether, dichloromethane, ethanol and other organic solvents, b.p. 77 ℃, the specific gravity is 0.894-0.898.

Special Instructions

1. In the process of using the Soxhlet extractor, use filter paper to pack the tea powder tightly to prevent the tea powder from leaking out and blocking the siphon.

2. The size of the filter paper straw should be appropriate. It can not only be tightly attached to the tube wall, but also absorb water. The height cannot extend to the height of the top liquid and the branch of the rising device, and it cannot be taken from the suction port of the siphon.

3. The liquid in the flask cannot be filled too much, generally 3-4 times the solvent of the Soxhlet extractor.

4. In the case of extraction, the quality of the sublimation operation is the key to the success of this experiment. The natural temperature during the sublimation process is too high, otherwise the paper will be carbonized and blackened, and some colored substances will be baked out to make the product not pure. In the second sublimation,

the firepower cannot be involved, otherwise a large amount of smoke will be inhaled, which will cause loss.

Suggested Waste Disposal

The extracted caffeine is finally recycled. The organic waste liquid generated during the experiment should not be poured into the drain, but should be poured into a special waste liquid treatment container. Inorganic acid waste liquid should also be poured into a special centralized treatment container.

Questions

1. What role does quicklime play in Method 1? What is the purpose of adding lead acetate solution in Method 2?

2. What is the difference between suction filtration and ordinary filtration, and what are its characteristics?

3. In the structure of caffeine, which nitrogen is the most alkaline? Please explain it.

4. What are the advantages and disadvantages of sublimation method?

Further Reading

Caffeine

4.2 Extraction of Berberine from Coptis Chinensis

Objectives

1. Learn the principles and methods of extracting berberine from coptis.

2. Further master the use of Soxhlet extractor.

Principles

Berberine is an alkaloid isolated from the Chinese herbal medicine Coptis, etc. It has a good effect on acute conjunctivitis, aphthous, acute bacillary dysentery and acute gastroenteritis. The content of berberine in berberine is about 4%-10%. In addition, three needles, barberry, celandine, celandine and other Chinese herbal medicines also contain berberine, but coptis and phellodendron have the highest content. There are three tautomers of berberine, as shown in Figure 4.2.1. From left to right, they are quaternary ammonium type, alcohol amine type and aldehyde type. In nature, they mostly exist in the form of quaternary ammonium bases.

Berberine is a yellow needle-like crystal containing 5.5 molecules of crystal water. After drying at 100 ℃, it loses molecular crystal water and turns to brownish red. Berberine is slightly soluble in water and ethanol, more soluble in hot water and hot ethanol, and almost insoluble in ether. Berberine hydrochloride, hydroiodide, sulfate, nitrate, are hardly soluble in cold water, but easily soluble in hot water. Therefore, it can be recrystallized with water to achieve the purpose of purification.

Apparatus and Reagents

Apparatus: Soxhlet extractor, Vacuum distillation device, Mortar, Water bath, Electric furnace, Oven, Büchner funnel, Suction filter flask, Separatory funnel, Beaker, Graduated cylinder, Filter paper, Absorbent cotton, Boiling stone.

Reagents: Coptis, 1% Acetic acid solution, Acetone (A.R), Saturated lime water, Concentrated hydrochloric acid (A.R), 0.5% Sulfuric acid solution, 95% Ethanol, Ice water.

Procedures

1. Extraction of berberine

(1) Weigh 10 g of coptidis, mash and grind and put it into the filter paper sleeve of the soxhlet extractor. Add 100 mL of 95% ethanol and a few boiling stones into the flask of the extractor. Install the extraction device of the extractor (Figure 4.1.1), connect the condensed water, heat to reflux, and continuously extract for 1-1.5 h (when the color of the extract is very light, stop the extraction). When the condensate is just siphoned down, immediately stop heating and cooling.

(2) Vacuum distillation. Recover most of the ethanol, until the remaining liquid in the bottle is brown-red syrup, stop the distillation.

(3) Add 30 mL of 1% acetic acid solution to the concentrated organic phase. After heating to dissolve, suction filter while hot to remove solid impurities. Add concentrated hydrochloric acid to the filtrate until the solution becomes turbid (about 10 mL).

(4) Cool the above solution in an ice-water bath. When it is lowered to room temperature, yellow needle-like crystals will precipitate out. Filter by suction. The obtained crystalline solid is washed twice with ice water and once with acetone to obtain the berberine hydrochloride crude product.

(5) Put the crude product into a 100 mL beaker, heat the water until it just

dissolves and boil, adjust the pH to 8.5-9.8 with lime milk. Cool and filter out impurities, continue to cool to below room temperature to precipitate berberine crystals, suction and filter to obtain yellow berberine crystals, dry in an oven at 50-60 ℃, and weigh the product.

2. Identification of berberine

(1) Take a little amount of berberine product, add 2 mL of concentrated sulfuric acid, add a few drops of concentrated nitric acid after dissolving. If the solution is cherry red, it means that berberine is present.

(2) Take about 50 mg of berberine finished product, add 5 mL of distilled water, slowly heat, add 2 drops of 20% sodium hydroxide solution after dissolution, it turns orange. Filter after cooling, add 4 drops of acetone to the filtrate, it will become turbid. A yellow precipitate of berberine was formed, indicating the presence of berberine.

(3) Take a small amount of berberine, dissolve it in water, and add a few drops of concentrated nitric acid. If the solution produces yellow-green berberine nitrate precipitation, it indicates the presence of berberine.

Notes

1. Before extracting berberine, coptis should be chopped and ground into powder, otherwise the extraction rate will be reduced.

2. The filter paper cartridge should be close to the wall of the device and be easily accessible. The height of the extract should not exceed the siphon, otherwise the extract cannot be fully soaked by the solvent, which will affect the extraction effect. The extracted material should not leak out of the filter paper tube, so as not to block the siphon tube.

3. During vacuum distillation, the temperature should not be too high, otherwise the product will be distilled out along with the ethanol, reducing the yield.

4. Berberine crystals should be dried at 50-60 ℃. If the temperature is too high, the berberine will deteriorate or become carbonized.

5. Berberine is oxidized by oxidizing agents such as nitric acid and will be converted into fuchsia berberine. In strong alkali, berberine will be partially converted into berberine aldehyde. Under these conditions, add a few drops of acetone. Condensation reaction occurs, and a yellow precipitate formed by the condensation of acetone and berberine aldehyde is formed.

Special Instructions

1. After the vacuum distillation is over, the heat source should be turned off first,

and the vacuum should be slowly released after a little bit of cooling. After the pressure inside and outside the system is balanced, the vacuum pump should be turned off to prevent back suction.

2. In the recrystallization process, the solvent cannot be too much otherwise it will cause product loss.

Suggested Waste Disposal

The berberine, the product of this experiment, was recovered. The organic waste liquid generated during the experiment should be poured into the container designated for organic wastes. Inorganic acid waste liquid cannot be poured directly into the drain, but should be poured into a special centralized treatment container.

Questions

1. If you want to recrystallize the crude berberine hydrochloride product, which solvent should be used?

2. In the process of extracting berberine from Coptis, what is the purpose of adjusting pH with lime milk?

4.3 Separation of the Pigments from Spinach Leaf

Objective

To extract photosynthetic pigments and separate them by column chromatography and thin-layer chromatography techniques.

Separation of the pigments from spinach leaf(operation video)

Principles

Green plants such as spinach leaves contain many natural products such as chlorophyll, carotene, water-soluble vitamins, lutein, etc. Spinach leaves present chlorophyll and β-carotene, these being primarily responsible for the leaf color, together with minor amounts of xanthophyll components.

Chlorophyll is green, and there are two similar structural forms, namely chlorophyll a ($C_{55}H_{72}N_4O_5Mg$) and chlorophyll b ($C_{55}H_{70}O_6N_4Mg$), both of which are complexes of porphyrins (substituted cyclotetrapyrrole-porphin derivatives) compounds and metals Magnesium. And they are necessary catalysts for plant photosynthesis. The a-type has a methyl group, whereas b-type has a formyl group attached to the porphyrin ring. Mg^{2+} ion loss in chlorophyll leads to the formation of pheophytin (a and b). Both chlorophyll a and chlorophyll b are easily soluble in organic solvents such as ethanol, ether, acetone, chloroform, and are also easily soluble in non-polar organic solvents such as ether and petroleum ether due to the large hydrocarbon structure. Chlorophyll a is a blue-black crystal with the melting point of 150-153 ℃. Its ethanol solution is blue-green with deep red fluorescence. Chlorophyll b is a dark green powder with a melting point of 120-130 ℃. The ethanol solution is green or yellow-green with red fluorescence, and has optical activity. Chlorophyll can be used as a non-toxic colorant for food, cosmetics and medicine.

Carotenes ($C_{40}H_{56}$) are conjugated polyenes (tetraterpenes) with a long-chain structure. There are three isomers, namely α-, β- and γ-carotene. Among them, β-isomers are the most important. In green plants with a long growth period, the content of β-forms in the isomers are as high as 90%. The β-body has the physiological activity of vitamin A. In the organism, the β-body is oxidized by enzymes to form vitamin A. At present, β-body can be produced industrially, used as vitamin A, and can also be used as food coloring.

Lutein($C_{40}H_{56}O_2$) is an oxygen-containing derivative of carotene, and its content in green leaves is usually twice than that of carotene. Compared with carotene, lutein is more soluble in alcohol and less soluble in petroleum ether.

The structural formulas of chlorophyll a, chlorophyll b, α-carotene, lutein and β-carotene are shown in Figure 4.3.1.

Thin-Layer Chromatography (TLC) is an important and effective method for separating and identifying mixtures, detecting compound purity, and tracking reactions.

Column Chromatography (CC) is an important method to separate and purify organics.

In this experiment, the above-mentioned pigments were extracted from spinach leaves, and separated, identified and purified by thin-layer chromatography and column chromatography.

Apparatus and Reagents

Apparatus: Chromatography column (chromatographic column), 150 mL Erlenmeyer flask, 60 mL Plow-shaped separatory funnel, Small glass funnel, Glass rod, Dropper, Ordinary filter paper, Absorbent cotton, 250 mL Beaker, 100 mL Graduated cylinder, 25 mL Graduated cylinder, Balance, Iron stand, Chromatography cylinder, Glass plate (10 cm×3 cm), Capillary.

Reagents: n-Hexane, Acetone ethanol, Sodium chloride, Anhydrous sodium sulfate, Silica gel (100 mesh), Quartz sand.

Procedures

1. Sample preparation

Weigh approximately 2 g of spinach leaves. Wash the leaves with water, remove the veins, cut up with scissors, and put the pieces in a mortar together with 22 mL of acetone, and 3 mL of hexane using a spatula (to prevent degradation of photosynthetic pigments). Grind the mixture until the leaves become discolored and the solvent takes on a deep-green color. Transfer only the resulting liquid to a separatory funnel (prevent the transfer of solids), and add 20 mL of hexane and 20 mL of aqueous solution of NaCl (10 wt%). Stir the mixture and decant. Discard the aqueous layer (bottom), wash the organic layer (top) with 20 mL of 10% NaCl solution and transfer the organic layer to a 50 mL Erlenmeyer flask. Dry the solution with anhydrous Na_2SO_4 and remove the desiccant by gravity filtration. Divide the sample into two parts. Take 1 mL of solution and reserve in a test tube for the analysis by TLC, and then use the remaining sample for CC separation.

2. TLC analysis of spinach pigments

Prepare a solution [hexane/aceton=7 : 3(V/V)], and place the eluent inside the chromatography tank (to about 5 mm high). Cover the cylinder cover and shake to make it saturated with solvent vapor.

With a pencil (use only graphite pencils) and a ruler, draw a horizontal line about 5 mm from the base of a TLC-foil and another horizontal line at a proper distance from the other end to the side (about 2 cm) as the eluent front (finish line).

With the help of a glass capillary, add a drop on the line at the TLC-foil. If the color of the spot is light, you can repeat the sample after the solvent evaporates.

After the spotting solvent evaporates, place the TLC plate with the spotting end down in the chromatography cylinder (dipped into the developing agent about 0.5 cm below the surface). Cover the cylinder cover and unfold after standing still. Pay attention to the elution process. When the front edge of the developing agent moves up to the upper end finish line, take it out immediately.

Wait for the solvent to evaporate, observe carefully and draw each spot with a pencil, measure and record the distance (measured to the center of the spot) of each spot. Calculate its R_f respectively.

3. Chromatographic separation of pigments by column chromatography

Put cotton or glass wool in the bottom of the column and fill about 10 mL eluent [hexane/ethanol=20 : 1(V/V)]. Slowly put white sand into the column until there is a 5 mm layer of sand over the cotton.

Weigh 20 g of silica gel on a 100 ml Erlenmeyer flask, and add approximately 50 mL of eluent [hexane/ethanol=20 : 1(V/V)], stirring with a glass rod or spatula, to form a slurry. Fill the column with the slurry, opening the column stop cock until the liquid is at the right edge of the stationary phase (it is very important for the column never to empty, completely or partially, and not to form air bubbles), collecting the solvent in a container for reuse. Sand can be added to the silica gel to form a layer of 1 or 2 cm thick on the edge of the stationary phase.

After the eluent in the column is discharged slightly below the surface of the sand, 5 mL of spinach leaf extract could be added to the chromatographic column. Open the chromatographic column stop cock to make the liquid sample enter the quartz sand layer, and then add a small amount of eluent to wash off the sample on the wall of column. After the sample enters the quartz sand layer, add the eluent for leaching. In order to save eluent, gradiant elution is carried out, 80 mL hexane-ethanol [20 : 1(V/V)], 40 mL hexane-ethanol [10 : 1(V/V)], 25 mL hexane-ethanol [5 : 1(V/V)] can be successively used in this experiment. Wash until all the ribbons are unfolded, and collect different ribbons separately to obtain the different pigments contained in the spinach leaves.

Notes

1. This line may not be drawn.
2. The nozzle of the capillary should be flat.
3. Do not damage the silica gel layer when spotting.
4. Do not immerse the sample point below the liquid level of the developing agent.
5. If the finish line is not drawn, take it out and draw the unfolding front.
6. TLC separation of spinach leaf pigments can generally present 7 spots of four colors, which are carotene (orangetin), pheophytin (gray), chlorophyll a and chlorophyll b (blue-green and yellow-green, 2 spots), and flavin (yellow, 3 spots). There are also cases where 8, 9 or even 10 spots are observed [Developing agent : n • hexane-ethand 7 : 3(V/V),GF silica gel plate]as shown in Table 4.3.1.

Table 4.3.1 **R_f values and color of different pigments in the spinach leaf**

Pigment	Color	R_f
Carotene	Yellow-orange	0.93
Pheophytin a	Gray	0.55
Pheophytin b	Gray(not visible)	0.47-0.54
Chlorophyll a	Blue-green	0.46
Chlorophyll b	Green	0.42
Xanthophylls	Yellow	0.41
	Yellow	0.31
	Yellow	0.17

Note: Eluent ∶ hexane/acetone=7 ∶ 3(V/V), GF silica gel plate.

Special Instructions

1. The silica gel layer should not be damaged when marking with a pencil.

2. Do not overload the thin-layer plate! An excessive amount of compound will cause overloading and result in large spots with considerable trailing. The spots from compounds with similar R_f values will overlap and correct analysis will be very difficult.

3. In the column chromatography separation, the height of the eluent in the column should never be lower than the top end of the silica gel.

Suggested Waste Disposal

Dispose of any leftover developing solvent and eluent in the recovery container for organic solvent after experiment.

Questions

1. When the spotted sheet is unfolded, why can't the spotted spot be immersed below the liquid level of the spreading agent?

2. Why does carotene move the fastest and have the largest R_f value?

Further Reading

The chemistry of vision——An important isomerization reaction

4.4 Analysis of Unknown Solutions of Alcohol, Phenol, Aldehyde, Ketone and Carboxylic Acid

Objectives

1. Review and master the chemical properties of alcohols, phenols, aldehydes, ketones and carboxylic acids.

2. Learn and master how to identify alcohols, aldehydes, ketones and carboxylic acids by chemical reagents.

Principles

Functional groups are atoms or groups of atoms attached to an organic compound that impart characteristic shapes and chemical properties to the compound. Because each functional group has a characteristic set of reactions, those reactions can be used to determine what functional groups actually are in the molecule. If these chemical reactions are applied to analyze and identify organic compounds, the following requirements shall be met: (1) Fast reaction speed. (2) There are obvious chemical phenomena, such as a specific color change, dissolution or formation of precipitation, release gas, etc. (3) High sensitivity. (4) High reaction specificity (refers to the specific reaction between a reagent and a functional group).

In this experiment, students are required to design a process to identify specified compounds according to the different chemical properties of each compound based on the reagents provided by the laboratory. Write out the plan and conduct the test after it is reviewed by the instructor.

Apparatus and Reagents

Apparatus: Beaker, Test tube, Test tube rack, Test tube clamp, Alcohol burner, Water bath kettle.

Reagents: 5% $K_2Cr_2O_7$ solution, Concentrated H_2SO_4, $KMnO_4$ solution, 1% $FeCl_3$ solution, 5% $CuSO_4$ solution, Phenolphthalein solution, 10% NaOH solution, Tollen's reagent, Fehling's reagent, 5% 2, 4-Dinitrophenylhydrazine, Liquid I_2, Saturated bromine water, 5% $AgNO_3$ solution, 5% $NaHCO_3$ solution, Concentrated ammonia solution, Litmus test paper, Ninhydrin solution, Lucas reagent.

Design Contents

Identify the following compounds by chemical reagents:

1. 1-Butanol, 2-butanol, *tert*-butanol and ethylene glycol.

2. Phenol, benzyl alcohol, phenylalanine, benzoic acid, salicylic acid and benzaldehyde.

3. Formic acid, ethyl acetoacetate, ethanal, acetone and phenol.

4. Glucose, fructose, starch and sucrose.

Procedures

1. Designing the experimental scheme

Review the reactions of a series of specified compounds. Independently design the identification scheme of all specified compounds with the given reagent (it is required to write out the experimental steps, expected phenomena, explanation and the relevant reaction equations).

2. Conducting identification experiments

During the experiment, carefully observe, record, correctly analyze and judge the experimental phenomena. Finally, complete the experiment report.

Notes

1. Many organic alcohols, phenols and ethers are toxic, and all are flammable. Acetone is highly flammable. Use these chemicals only in well-ventilated space. Keep away from flames and other sources of ignition.

2. Bromine is corrosive which causes serious burns. Take great care to avoid contact with skin, eyes, and clothing. In case of accidental contact, flood the affected area with copious amounts of water and seek medical attention.

3. After a long-term storage, Tollen's reagent will generate AgN_3 precipitation which is easy to explode, so it needs to be prepared temporarily. During the experiment, do not directly heat with flame. After the experiment, add a little dilute nitric acid, boil and wash the silver mirror immediately.

4. Chromium is highly toxic and its acid solution is extremely corrosive. Avoid ingestion. Handle only with gloves. Take great care to avoid contact with skin, eyes, and clothing. In case of accidental contact, flood the affected area with copious amounts of water. In case of ingestion, seek medical attention immediately.

5. Silver nitrate is highly oxidizing. Upon contact with silver nitrate, skin will turn dark. Take great care to avoid contact with skin.

Suggested Waste Disposal

Discard all wastes in the appropriately labeled containers. No organic materials from this experiment are allowed to poured into the drain!

Questions

1. Point out as many methods as possible to identify aldehydes and ketones.

2. There are 5 bottles of unlabeled reagents, namely formic acid, benzoic acid, benzaldehyde, phenol and benzyl alcohol. Try to select appropriate reagents to identify them.

3. Combined with the experimental results, analyze the key to the success or failure of the experiment. What needs to be improved in this experiment?

4.5 Synthesis of Biodiesel

Objectives

1. Learn the principles and methods of biodiesel synthesis.

2. Understand biomass energy and green energy.

Principles

The global mineral energy reserves are very limited, but the world's consumption of mineral energy is increasing. The development of renewable and environmentally friendly alternative fuels has become one of the most important issues for sustainable economic development, and the use of biofuel technology has emerged. As a clean biofuel that can replace petrochemical diesel, biodiesel is an environmentally friendly fuel with production cost and performance basically equivalent to petrochemical diesel, with good environmental characteristics and biodegradability, and has broad development prospects.

Biodiesel can be prepared by physical and chemical methods. Physical methods include direct mixing and microemulsion methods. Chemical methods include high-temperature thermal cracking and transesterification. The transesterification method is currently the main method for synthesizing biodiesel. Various natural vegetable oils and animal fats, as well as waste oils from the food industry, can be used as raw materials for transesterification to produce biodiesel.

In the transesterification reaction, the main component of the oil, triglyceride, and various short-chain alcohols undergo transesterification under the action of a catalyst to obtain fatty acid methyl esters and glycerol. Alcohols that can be used for transesterification include methanol, ethanol, propanol, butanol and pentyl alcohol. Alcohol, among which methanol is the most commonly used, is due to the low price of methanol, short carbon chain, strong polarity, can quickly react with fatty acid glycerides. And alkaline catalysts are easily soluble in methanol. The transesterification method includes acid catalysis, base catalysis, biological enzyme catalysis and supercritical transesterification. The equation of the transesterification reaction principle is shown as follows.

$$\begin{array}{l} H_2C—COOR' \\ \quad | \\ HC—COOR'' \\ \quad | \\ H_2C—COOR''' \end{array} + 3CH_3OH \xrightarrow{\text{catalyst}} \begin{array}{l} H_2C—OH \\ \quad | \\ HC—OH \\ \quad | \\ H_2C—OH \end{array} + \begin{array}{l} R''COOCH_3 \\ R'COOCH_3 \\ R'''COOCH_3 \end{array}$$

Apparatus and Reagents

Apparatus: Three-neck round bottom flask, Stirrer, Spherical condenser, Thermometer, Water bath, Iron stand, Separatory funnel.

Reagents: Vegetable oil, Potassium hydroxide, Methanol, Hydrochloric acid solution, Anhydrous sodium sulfate, Periodic acid, Potassium iodide, Sodium thiosulfate, 0.5% Starch indicator.

Procedures

1. Synthesis of biodiesel

(1) As shown in Figure 4.5.1, put 100 g of soybean oil into a 250 mL three-necked flask equipped with a condenser. After heating to 65 ℃, add 1.4 g of catalyst KOH and 21.8 g of methanol solution under stirring. Start timing after constant temperature to allow the mixture to react fully for 1 h Afterwards, the reaction mixture was taken out and placed in an ice-water bath to make the reaction complete in time.

(2) The reaction mixture is placed in a separatory funnel for static stratification. The upper layer is a yellow mixture of methyl ester (biodiesel) and methanol, and the lower layer is brown, which are glycerin and unreacted triglycerides. The upper liquid was collected and distilled at 70 ℃ under atmospheric pressure to separate methanol and methyl esters.

(3) The residue of distillation is washed once with 36% hydrochloric acid, and then washed with water at 85 ℃ for several times until there is no obvious milky white substance in the water phase, and the fatty acid salt, glycerin, water-soluble substance, free fatty acid, etc. are removed. After washing, add enough anhydrous sodium sulfate to the residue, shake well, stand for 10 min and then filter to remove the anhydrous sodium sulfate for drying. When the oil layer turns into a light yellow transparent liquid, remove the biodiesel. Weigh and calculate the transesterification rate.

2. Determination of glycerol content in synthetic products

(1) Use an electronic analytical balance to weigh 0.2 g and 0.4 g (accurate to 0.000 1 g) of biodiesel each in two beakers, add water to dissolve, and after cooling, transfer to

100 mL volumetric flasks, wash and set the volume for later use.

(2) Measure 25 mL of the prepared solution into 250 ml Erlenmeyer flasks ① and ②, measure 25 mL water into a 250 mL Erlenmeyer flask ③, and then add 20 mL 0.02 mol/L KIO_4 solution and 10 mL H_2SO_4 solution (3 mol /L), cap the bottle, shake well and place in a dark place at room temperature for 30 min.

(3) Then add 2 g KI, 150 mL water, and titrate the precipitated I_2 with the configured $Na_2S_2O_3$ standard solution. When the titration is light yellow, add 1 mL 0.5% starch indicator and continue the titration until the blue color just disappears. Do a parallel test and blank test.

Notes

1. The catalyst used in the acid catalysis method is an acid catalyst, mainly sulfuric acid, hydrochloric acid and phosphoric acid. Under the conditions of the acid catalysis method, free fatty acids will undergo an esterification reaction, and the esterification reaction rate is much faster than the transesterification rate, so this method is suitable for the preparation of biodiesel from free fatty acids and oils with high moisture content, and its yield is high. However, the reaction temperature and pressure are high, the amount of methanol is large, the reaction speed is slow, and the reaction equipment needs stainless steel materials. Industrially, acid catalysis has received much less attention than alkali catalysis.

2. The yield of biodiesel produced from vegetable oil is:

Transesterification rate = mass of methyl ester ÷ mass of rape oil added.

3. The reaction process for determining the glycerin content is as follows.

Oxidation:
$$\begin{array}{l} H_2C\text{—}OH \\ \quad | \\ HC\text{—}OH \\ \quad | \\ H_2C\text{—}OH \end{array} + 2HIO_4 \longrightarrow 2HCHO + HCOOH + 2HIO_3 + H_2O$$

Reduction:
$$HIO_4 + 7I^- + 7H^+ = 4I_2 + 4H_2O$$
$$HIO_2 + 5I^- + 5H^+ = 3I_2 + 3H_2O$$

Titration:
$$I_2 + 2Na_2S_2O_3 = 2NaI + Na_2S_4O_6$$

Special Instructions

1. During the distillation process, the mercury bulb of the thermometer should be at the same level as the lower port of the branch pipe, and the heating process is not easy to be too fast.

2. When drying the product with anhydrous sodium sulfate, the drying should be done so that the desiccant neither sticks to the wall nor sticks into a block. At the same

time, the desiccant should not be added too much, otherwise it will cause the product to be adsorbed and cause loss.

3. When configuring the standard solution, pay attention to the correct use of the volumetric flask; when the volume is constant, the line of sight should be flush with the scale line.

4. When titrating, the tip of the burette should be inserted 1-2 cm below the mouth of the conical flask (or the mouth of the beaker), the titration speed should not be too fast, 3-4 drops per second is appropriate, and it must not flow down into a liquid column. Shake while dripping. When approaching the end point, it should be added in one or half drop with slow dripping and quick stirring. The end point of the titration is that the color does not change for 30 seconds.

Suggested Waste Disposal

The remaining triglycerides and the produced biodiesel are recycled separately, and the waste liquid generated during the experiment is poured into a special waste liquid treatment container, and must not be poured into the drain. Inorganic acid waste liquid should also be recycled into the corresponding treatment container, and should not be poured directly into the drain.

Questions

1. According to the experimental results, how to calculate the glycerin content in the product?

2. Why is KOH usually used as the base catalyst in the base-catalyzed transesterification method?

4.6 Preparation of Methyl Orange

Objectives

1. Master the experimental operation of diazotization reaction and coupling reaction.

2. Consolidate the experimental operations of salting out and recrystallization.

Principles

Methyl orange (sodium *p*-dimethylamino azobenzene sulfonate), also known as Golden Lotus Orange D, is widely used in production and scientific experiments. It is often used as an acid-base indicator and biological dye. It can also be used for printing and dyeing textiles and spectrophotometric determination of chlorine, bromine and bromide ions. Methyl orange is obtained by coupling diazonium *p*-sulfanilic acid salt and

N,N-dimethylaniline acetate in a weakly acidic medium. The first result of the coupling is a bright red acid methyl orange. It is acid yellow, which turns into orange sodium salt in alkali, namely methyl orange, as shown in Figure 4.6.1. Methyl orange is slightly soluble in water and appears yellow, soluble in hot water, and almost insoluble in ethanol.

Apparatus and Reagents

Apparatus: Beaker, Test tube, Bühner funnel, Erlenmeyer flask, Filter flask, Glass rod.

Reagents: Sulfanilic acid crystals, 5% Sodium hydroxide solution, Sodium nitrite, Concentrated hydrochloric acid, N, N-xylidine, Glacial acetic acid, Ethanol, Ether, Starch-potassium iodide test paper, pH Test paper, Ice water.

Procedures

1. Preparation of diazonium salt

(1) Place 2.0 g (0.0115 mol) of *p*-aminobenzenesulfonic acid crystals in a 100 mL beaker, add 10 mL of 5% sodium hydroxide solution, dissolve it with warmth, and cool to room temperature. Dissolve 0.8 g sodium nitrite (0.0116 mol) in 6 mL water, add the above solution, and cool to 0-5 ℃ in an ice-salt bath.

(2) Add 3 mL concentrated hydrochloric acid and 10 mL water to dilute hydrochloric acid solution, slowly drip into the cooled mixed solution, stir while dripping, keep the temperature below 5 ℃, after the dripping, use starch-potassium iodide test paper inspection, the test paper should be blue (if it does not show blue, add sodium nitrite solution), continue to stir in the ice bath for 15 min, you can observe the precipitation of white fine grained diazonium salt.

2. Coupling reaction

(1) Mix 1.3 mL (1.248 g, 0.0103 mol) N,N-xylene amine and 1 mL glacial acetic acid in a test tube, slowly add to the prepared diazonium salt cold suspension, continue after dropping. Stir in an ice bath for 10 min to complete the coupling. During this process, a red precipitate can be observed.

(2) Under stirring, slowly add 5% sodium hydroxide solution (about 25 mL) to the mixed solution until the pH test paper becomes alkaline, and the precipitation turns from red to orange. Then the reaction mixture was heated on a boiling water bath for 5 min, cooled to below room temperature, then cool with ice water to precipitate methyl orange crystal completely. The crystals were collected by suction filtration, washed with a small amount of water, then washed with a small amount of ethanol and

a small amount of ethyl ether, and then pressed to dryness to obtain a crude methyl orange product.

3. Purification

The crude methyl orange product is dissolved in dilute sodium hydroxide solution, heated to dissolve, cooled to crystallize, filtered by suction, washed with a small amount of ethanol and ether, and dried to obtain the finished product of methyl orange.

4. Identification of methyl orange

Take a small amount of methyl orange product and dissolve it in water. It should be orange. After adding dilute hydrochloric acid, the solution turns purple, and after adding dilute NaOH, the solution turns orange.

Notes

1. During the diazotization process, the temperature should be strictly controlled. If the reaction temperature is higher than 5 ℃, the resulting diazonium salt is easily hydrolyzed into phenol, reducing the yield.

2. The starch-potassium iodide test paper just turned blue, indicating that the nitrous acid has just been excessive and the diazotization is complete, otherwise add sodium nitrite aqueous solution. Use starch-potassium iodide test paper to test, if excessive nitrous acid will cause side reactions, urea should be added to remove.

3. When the sodium hydroxide solution is added dropwise until the solution turns orange, the addition can be stopped, because the unreacted N,N-dimethylaniline acetate salt can precipitate insoluble in water under alkaline conditions. N, N-dimethyl aniline affects the purity of the product.

4. Because the product is alkaline, and the temperature is high, it is easy to deteriorate, and the color becomes darker. Therefore, the recrystallization should be rapid. The reactant should not be heated in a water bath for too long and the temperature should not be too high. The purpose of washing with ethanol and ether is to dry the crystals quickly, and the color of the wet methyl orange will become darker quickly after being exposed to light in the air.

Special Instructions

1. The temperature control of this reaction is very important. When preparing diazonium salt, the temperature should be controlled below 5 ℃.

2. The heating time of the reaction product in the water bath should not be too long (about 5 min), and the temperature should not be too high (about 60-80℃), otherwise the product will deteriorate and the color of the product will become darker.

3. Because the product crystals are relatively fine, the filter paper should be prevented from being broken during suction filtration (the Buchner funnel does not need to be packed too tightly). Wet methyl orange will deteriorate even when exposed to sunlight. It is usually washed with ethanol or ether and dried at 55-78 ℃. The resulting product is a sodium salt with unfixed melting point.

Suggested Waste Disposal

To recycle product methyl orange crystals, organic waste liquid and inorganic waste liquid are separately poured into the corresponding waste liquid treatment container, and shall not be poured directly into the sewer. The used concentrated hydrochloric acid should also be sealed in time.

Questions

1. Why can diazonium salt be coupled with phenol or amine? What are the coupling conditions?

2. In this experiment, when preparing diazonium salt, why should *p*-aminobenzene sulfonic acid be turned into sodium salt? If this experiment is changed to the following operation steps, first mix *p*-aminobenzene sulfonic acid and hydrochloric acid, and then add sodium nitrite solution for diazotization, is it feasible?

3. Try to explain the reason for the discoloration of methyl orange in acidic medium, and express it with the reaction formula.

4. What is the diazotization reaction? What are the conditions for the diazotization reaction?

4.7 Integrating Computational Molecular Modeling into the Organic Chemistry Experiment

Integrating computational molecular modeling into the organic chemistry experiment

Organic chemistry experiments based on molecular simulation (experimental operation)

Chapter 5 Experimental Literature

Experimental literature

Appendix

Appendix Ⅰ Atomic Weights of Most Common Elements

The atomic weight of common elements are shown in Table A.1.1.

Table A.1.1 The atomic weight of common elements

Element	Symbol	Atomic weight/(amu, g/mol)	Element	Symbol	Atomic weight/(amu, g/mol)
Silver	Ag	107.868	Iodine	I	126.905
Aluminum	Al	26.982	Potassium	K	39.098
Bromine	Br	79.904	Magnesium	Mg	24.305
Carbon	C	12.011	Manganese	Mn	54.938
Calcium	Ca	40.078	Nitrogen	N	14.007
Chlorine	Cl	35.453	Sodium	Na	22.990
Chromium	Cr	51.996	Oxygen	O	15.999
Copper	Cu	63.546	Phosphorus	P	30.974
Fluorine	F	18.998	Lead	Pb	207.200
Iron	Fe	55.847	Sulfur	S	32.066
Hydrogen	H	1.008	Tin	Sn	118.710
Mercury	Hg	200.59	Zinc	Zn	65.390

Appendix Ⅱ Densities and Concentrations of Common Acid and Base Solutions

Densities and concentrations of common acid and base solutions

Appendix Ⅲ Preparation of Commonly Used Reagents

Preparation of commonly used reagents

Appendix Ⅳ Physical-Chemical Properties of Common Organic Chemicals

Physical-chemical properties of common organic chemicals